Affordable Housing in the U

Affordable Housing in the United States addresses the issue of affordability of housing, or the lack thereof, going beyond conventional policy discussions to consider fundamental questions such as: What makes housing affordable and for whom is it affordable? What are the consequences of a lack of affordable housing? How is affordable housing created? And what steps can be taken to ensure all people have access to affordable housing?

With the understanding that different households face different challenges, the book begins by breaking down the variables relevant to the study of affordable housing, including housing costs, household income, geographic location, and market forces, to help readers understand and quantify affordability at the individual and societal level. Part II examines the consequences of unaffordable housing, highlighting racial inequities in housing access and affordability, and multiple forms of housing precarity including eviction and homelessness. Part III explores the entities involved in providing affordable housing such as local and federal governments, regulatory agencies, non-profit organizations, and for-profit developers. In Part IV, case studies from U.S. cities demonstrate the complex web of organizations, policies, and market conditions that influence housing affordability, revealing substantial regional variations in access and policy responses. Part V proposes a future roadmap and outlines four potential states with radically different outcomes for the affordable housing system in the United States.

An ideal book for graduate and undergraduate courses in economics, public policy, real estate finance and development, sociology, and urban planning, this title will also be of value to professionals and policymakers seeking to understand and improve housing affordability and access.

Gregg Colburn is an Associate Professor in the Runstad Department of Real Estate, College of Built Environments, at the University of Washington. He teaches courses in housing, economics, and finance at both the undergraduate and graduate levels. His research focuses on housing policy, housing markets, housing affordability, and homelessness. He is also actively engaged in community efforts to address the acute housing crisis in the Puget Sound region. He is the author of *Homelessness is a Housing Problem* (2002).

Rebecca J. Walter is an Associate Professor in the Runstad Department of Real Estate, College of Built Environments, at the University of Washington. Her research is focused on policy innovation in low-income housing. She emphasizes a spatial analytical approach to examine how housing policies either expand opportunity or perpetuate inequality for low-income households. Most of her work is applied as it involves direct engagement with public housing authorities and non-profit housing providers.

Affordable Housing in the United States

Gregg Colburn and Rebecca J. Walter

NEW YORK AND LONDON

Designed cover image: by ah_fotobox – Andreas*H/Getty Images

First published 2025
by Routledge
605 Third Avenue, New York, NY 10158

and by Routledge
4 Park Square, Milton Park, Abingdon, Oxon, OX14 4RN

Routledge is an imprint of the Taylor & Francis Group, an informa business

© 2025 Gregg Colburn and Rebecca J. Walter

The right of Gregg Colburn and Rebecca J. Walter to be identified as authors of this work has been asserted in accordance with sections 77 and 78 of the Copyright, Designs and Patents Act 1988.

All rights reserved. No part of this book may be reprinted or reproduced or utilised in any form or by any electronic, mechanical, or other means, now known or hereafter invented, including photocopying and recording, or in any information storage or retrieval system, without permission in writing from the publishers.

Trademark notice: Product or corporate names may be trademarks or registered trademarks, and are used only for identification and explanation without intent to infringe.

ISBN: 978-1-032-41169-9 (hbk)
ISBN: 978-1-032-40726-5 (pbk)
ISBN: 978-1-003-35658-5 (ebk)

DOI: 10.1201/9781003356585

Typeset in Times New Roman
by KnowledgeWorks Global Ltd.

Contents

Book Summary vii
Acknowledgments viii

PART I
Introduction 1

  1 What is Affordable Housing? 3

PART II
Understanding Affordable Housing 13

  2 The Housing Stock in the United States 15

  3 Deconstructing Affordability 26

  4 Historical Perspectives on Affordable Housing 37

  5 Race and Affordable Housing 54

  6 Housing Instability 66

PART III
Providing Access to Affordable Housing 81

  7 Sectors and Actors Involved in the Provision of Affordable Housing 83

  8 Supply Side Housing Assistance 95

  9 Demand Side Housing Assistance 112

10 Affordable Homeownership   125

11 Regulatory Strategies   141

PART IV
Case Studies   **157**

12 Chicago, IL   159

13 San Antonio, TX   169

14 Seattle, WA   179

PART V
The Path Forward   **191**

15 An Affordable Housing Roadmap   193

Index   *207*

Book Summary

The lack of affordable housing in the United States is a growing crisis with a long and complex history. In *Affordable Housing in the United States*, Gregg Colburn and Rebecca Walter provide a book exclusively focused on the somewhat vague notion of affordable housing. To bring clarity to a wide audience of potential readers, the authors go beyond conventional discussions of housing policy and consider a broader list of fundamental questions: What makes housing affordable and for whom is it affordable? How is affordable housing created or preserved? What role do market forces play in the provision (or lack thereof) of housing that is affordable? How do federal, state, and local governments support or promote affordability? What are the consequences of unaffordable housing? And, what steps can and should be taken to ensure that all people have access to housing that is affordable to them? The purpose of the book is to provide readers with a thorough review of the multiple domains and forces that interact to constitute the field of affordable housing.

Acknowledgments

We would like to express our gratitude to all individuals who generously dedicated their time and expertise to review the chapters in this book. Their invaluable feedback and constructive criticism have enhanced the quality and accuracy of the content. Our gratitude to Arthur Acolin, Ryan Allen, Andrew Aurand, Steven C. Bourassa, Katrina Carrasco, Allison Clements, Sara Eaves, Chris Hess, Andria Lazaga, Keith Leung, Alex Ramiller, Eric Seymour, and Ruoniu (Vince) Wang. We would especially like to thank two graduate students, Amber Khan and Amy Youngbloom, for their work to locate and process data for many of the figures and tables in the book, as well as background work to better understand the literature in this field. Their dedication, attention to detail, and hard work have been invaluable in ensuring the success of this project. We would also like to thank two undergraduate students, Marie Pino and Junxi (Cissi) Xu, for their assistance to locate and process preliminary data for the figures and tables. Finally, we would like to thank the editorial team at Routledge for their support as we prepared, and finalized, our manuscript.

Part I

Introduction

Chapter 1

What is Affordable Housing?

Overview

"What is affordable housing?" If someone were to stop you on the street and ask this seemingly simple question, how would you respond? If you aren't able to come up with a clear and concise answer, you are not alone. Most people have an intuitive sense of what housing is affordable to themselves, and what housing isn't. But when thinking about the population as a whole, it is harder to define what constitutes affordable housing. This complexity may partly explain why responses to the affordable housing crisis are frequently inadequate, and, at times, misguided. For Elon Musk and Bill Gates, *all* housing is affordable, while for workers who earn the minimum wage, only a small subset of the existing housing stock is. Therein lies the challenge of affordable housing: what's affordable for one person is unaffordable to the next. Complicating matters is the fact that programs and policies designed to promote housing affordability may define the concept of affordability differently. To provide clarity for a broad range of readers interested in this topic, our book seeks to answer a range of important questions: How do we define affordable housing? Why do some communities have a sufficient supply of affordable housing while others face severe shortages? What policies and programs are used to develop or preserve affordable housing? Why is ensuring that all households have access to affordable housing an essential societal goal? How can our nation, states, and localities ensure that all households have access to housing that is affordable based on their own financial and personal circumstances?

Books about the field of housing cover a wide range of topics beyond affordability; we recommend that readers who want a broader understanding of the field consult these works.[1] Here, we will focus specifically on affordable housing with the goal of providing a consolidated guide to the topic. This book includes discussions of housing affordability definitions, housing development (how is housing constructed and maintained), housing markets (how are prices for housing established), housing policy (what policies and regulations impact the provision of affordable housing), and the variety of social consequences that arise when households cannot procure housing that is affordable. We hope that this volume serves as a comprehensive resource for people teaching courses on affordable housing, for policymakers who are increasingly tasked with responding to the affordability crisis, and for advocates seeking a roadmap for a future that ensures affordable housing for all.

Both authors of this book (Rebecca and Gregg) teach courses related to housing at the University of Washington in Seattle. As we wrote the first draft of this book in 2022 and 2023, evidence of a nationwide affordable housing crisis was mounting. National and local media outlets were filled with stories of rapidly accelerating home values, skyrocketing rents, bidding wars for homes, lines of people waiting to tour an apartment for rent, and the increasing prevalence of evictions and homelessness as Covid-19 eviction moratoria were lifted. Housing affordability

DOI: 10.1201/9781003356585-2

has become one of the most common responses in polls that ask respondents to identify the public issues that are most important to their communities, replacing other popular concerns such as traffic and the economy. Housing now consumes a greater percentage of the household budget than it did decades ago for low- and middle-income households (Schanzenbach et al., 2016). Housing production has not kept up with population growth in many communities (Kingsella & MacArthur, 2022). There is a glaring lack of housing options for low-income households in every state in the nation (NLIHC, 2024), and a lack of affordable housing contributes to other social ills, including displacement and segregation (Solomon et al., 2019). As the crisis of housing (un)affordability becomes more acute, the need to understand the crisis and its solutions has never been more urgent.

A common misperception is that housing affordability is only an issue for residents of expensive coastal cities such as New York and San Francisco. While it is surely the case that housing costs are a major concern for residents of these cities, the lack of affordable housing is an issue in every corner of the country. In fact, there is no state in the nation where there is a sufficient supply of affordable housing for households with the lowest incomes. According to data from the National Low Income Housing Coalition, there is a shortage of 7.3 million rental units available to extremely low-income renters in the U.S.,[2] and for every 100 extremely low-income renter households, only 34 units are available and affordable (NLIHC, 2024). This evidence underscores that affordable housing is a standing concern in all communities. The recent increase in focus and attention on housing-related issues from all levels of government highlights the growing realization of the need for fundamental changes to our housing system in this country.[3]

The Rent is Too Damn High

In 2005, Jimmy McMillan, of Brooklyn, New York, founded the Rent is Too Damn High Party. McMillan was the party's nominee for New York City mayor in 2005 and 2009, and also ran for governor of New York in 2010. While unsuccessful in each of these attempts, McMillan highlighted the significant challenge posed by unaffordable housing in New York City. McMillan's distinctive political style and appearance received considerable public attention and even garnered a parody on Saturday Night Live's Weekend Update segment. While unsuccessful in his attempts to win election to public office, McMillan's message resonated with millions of New Yorkers who struggle with housing costs.

Beyond the borders of New York City, many readers of this book might also share the perspective that the rent is, indeed, too high when compared to the cost of housing in prior decades. Empirical evidence suggests that their hunch is correct. Since the 1970s, inflation-adjusted rental housing costs have grown by nearly 20%, while inflation-adjusted household income has fallen by over 10%. As a result, housing costs now consume a far greater percentage of the household budget than 50 years ago. Given the public concern associated with housing affordability, measuring and quantifying this concept has become an important analytical endeavor. Figure 1.1 demonstrates how housing costs and incomes have diverged over the past half century.

A lack of affordable housing has numerous adverse consequences. Obviously, homelessness represents the most acute consequence of unaffordable housing. The inability to procure housing that is affordable and accessible leads to housing precarity and, for some, the crisis of homelessness. Research demonstrates that cities with high housing costs and low vacancy rates have disproportionately high rates of homelessness; high-cost coastal cities are where those conditions are most prevalent (Colburn & Aldern, 2022). Beyond homelessness, there are many negative effects of a lack of affordable housing. Spending too much on rent crowds out other

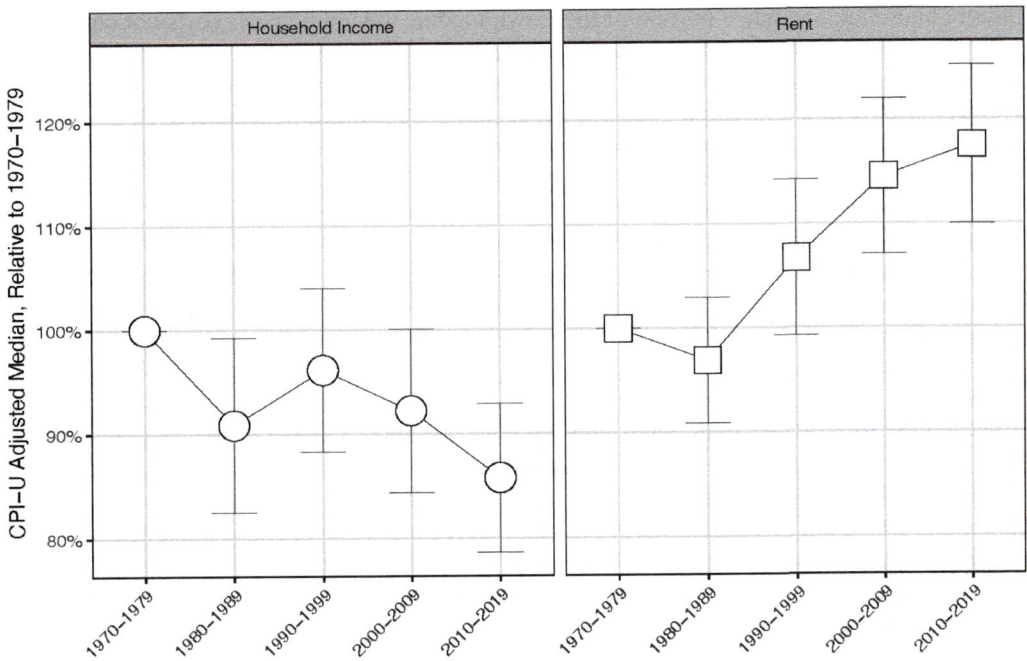

Figure 1.1 Inflation-adjusted Median Household Income and Rent for Renter Households in the U.S., 1970–2019
(Source: Panel Study of Income Dynamics; Colburn et al., 2024)

necessary household expenditures, leads to reduced health outcomes for children, negatively impacts mental health, and can even lead to mortality (Airgood-Obrycki et al., 2022; Aratani et al., 2011; Bentley et al., 2019; Elliott et al., 2021; Graetz et al., 2024). More broadly, the crisis of a lack of affordable housing is changing the way that communities operate. Essential professionals struggle to find housing that is affordable to them in many expensive locations. School teachers, firefighters, and hospitality workers, for instance, are essential to the success and viability of a community, but in many locations, the wages provided for these careers are inadequate to obtain housing in the communities that they serve. This creates an imbalance in local labor markets with more lower wage jobs available than affordable rental units (Zonta, 2020). These trends highlight the importance of this topic and underscore the importance of affordable housing in the long-term health of communities.

Measuring Housing Affordability

It is likely clear to many readers that affordability is a relative, not absolute, concept. We think about housing affordability relative to the resources that a household must pay for such housing; and most commonly, we think about housing costs relative to household income. But what percent of a household's budget is reasonable to devote to housing? Even that question is difficult to answer, but consensus has centered on 30%. How did we arrive at a 30% threshold to determine housing that is affordable to a household? That measure is enshrined in public policy. Because many federal housing programs, such as public housing and housing vouchers, require households to pay a portion of their own income toward housing costs not covered by the subsidy or support, the government had to determine the amount that households would be required to pay. In 1940, the original cap for housing affordability was set by the federal government at 20% of a household's income for federally subsidized housing (HUD, 2017). This threshold lasted for nearly thirty years, but changed in 1969 pursuant to the Brooke Amendment, when it was determined that residents of public housing would need to pay no more than 25% of their household income toward rent for the units that they occupied (HUD, 2014). The threshold was again raised in 1981 to 30% where it now stands (HUD, 2017). In addition to serving as an important measure for federal housing policy, the 30% threshold has been adopted for other uses as well. In particular, that threshold is now used by researchers as the primary measure of determining whether a household's consumption of housing is "affordable" or not.

There are a range of factors that complicate efforts to establish clear measures of housing affordability. First, households with high incomes are able to spend a greater percentage of their income toward housing without constraining other household consumption. A household earning a million dollars a year could spend over 30% of its income toward housing, but that household would not be constrained in any reasonable sense by living in housing that is deemed *unaffordable*. Contrast that household with one that earns $35,000 a year; housing that consumes a third or half of its income has clear, and damaging, implications for that household's ability to afford other household necessities.

A second complicating factor is whether a household rents or owns its housing. Housing affordability measures are easier to understand when a household rents its housing. Some portion of the household budget is used to make rental payments to a landlord and to pay for household utilities. The sum of these two amounts determines whether those payments exceed the 30% threshold for affordability. In situations where households own a home, the analysis becomes more complicated. Rather than paying monthly rent, most homeowners finance their homes by taking out a mortgage. Each month, homeowners make payments to their lender in satisfaction of their mortgage debt; these payments cover both interest and principal payments. The most common mortgage structure is a 30-year fixed loan, paid monthly. Therefore, a homeowner using this type of financing would make the same monthly payment over the life of the loan (360 payments in total). At the end of the 30 years, all interest payments will have been made and the principal loan amount will have been paid off. On top of the mortgage payment, homeowners must also pay property taxes which must be included when calculating the costs of homeownership. Therefore, to determine the monthly housing costs of a homeowner, one could sum the monthly cost of a mortgage, monthly property tax payment, and monthly utilities. Dividing that sum by the household income would tell us whether that household is breaching the 30% threshold for affordability.

There are several problems with this approach. First, monthly mortgage payments include both principal payments and interest payments. Principal payments don't change the net worth of a household – they simply reduce the amount of outstanding mortgage. Therefore, it really isn't an expense in the truest sense of the word. The interest payments are clearly an expense as they are payments made to the bank. But even this portion of the payment is fraught with complexity. The federal government, in an effort to promote homeownership, has enacted numerous policies to make homeownership more attractive. One such policy is to make interest payments on a mortgage tax deductible. Therefore, the true cost of interest payments made to a bank is less than the stated amount for homeowners that are eligible for this tax deduction. As such, a household might have an incentive to take out a mortgage that would create a monthly payment that would be deemed unaffordable by the 30% threshold, but the household may be happy to do so because of tax benefits associated with those payments. A related benefit of homeowners is the tax deductibility of property tax payments for state and local taxes. Again, this feature masks the true costs of homeownership. As a result, using affordability measures for homeowners is a challenging exercise and is not comparable to renter households for the reasons outlined above.

The 30% housing affordability threshold has now been operationalized into a common measure of housing affordability called *housing cost burden*. This intuitive concept measures the percentage of a household's income that is devoted to housing costs (including insurance and utilities). The higher the percentage, the more burdened the household. Most commonly, a household is housing cost burdened once this fraction reaches 30%. Above 50%, households are considered severely cost burdened. While intuitive and easy to understand and calculate, the 30% measure is not a perfect measure of the burden of high housing costs. First, the break points are somewhat arbitrary. A household with a 31% housing cost burden is not in a materially different situation than a household with a burden of 29%, although only the former household would be considered housing cost burdened per the conventional definition. A low-income household with the same housing cost burden as a high-income household (say, 33%) faces a very different reality in meeting their monthly financial obligations. After devoting a third of their income to housing, the wealthy household has significant resources left over to pay other household expenses, while the low-income household does not (Herbert et al., 2018). One could argue that, in reality, the *burden* on the low-income household is far greater than on the high earning one, even though the housing cost fractions are identical. Housing cost burden also fails to capture a range of tradeoffs that households might make in order to achieve housing affordability, such as moving in with other households, living in smaller, or substandard housing, or residing in a less desirable neighborhood (Belsky et al., 2005; Leishman & Rowley, 2012).

Scholars and analysts have developed a range of competing measures of housing affordability to address the obvious shortcomings of the housing cost burden measure. A common method is known as the *residual approach* which measures the amount of money left over after housing costs are paid (Herbert et al., 2018; Stone, 2006). The residual method calculates the amount of income – post housing costs – that households of different sizes should have to meet basic needs. A shortcoming of this approach is the challenge of creating a universal measure of non-housing costs of living. A related measure is called *shelter poverty*. Using the residual method, shelter poverty identifies households that are unable to afford non-housing related household necessities (Stone, 1993). Researchers in Australia have developed the *30:40 Rule* or *Housing Affordability Stress*. This measure identifies only those households that are in the bottom 40% of the income distribution who spend over 30% of their household income on housing costs (AHURi, 2019). This measure deals with the thorny issue of high-income households that are housing cost burdened by choice.

A different approach is the *housing wage* which creates an estimate of the hourly wage that a full-time worker must earn in order to pay market rent – the typical amount a property will rent for without any restrictions – without spending more than 30% of income toward housing (Aurand et al., 2022). In 2022, the average national hourly wage was $25.82 for a modest two-bedroom apartment, and was $21.25 for a one-bedroom. The average renter's wage is $21.99 – suggesting that a significant percentage of renters are unable to afford basic rental housing (Aurand et al., 2022). Beyond these measures, researchers have also developed more complicated approaches that account for neighborhood and location factors when assessing affordability (Ezennia & Hoskara, 2019).

The federal government has established a separate category that measures housing affordability challenges known as the *Worst Case Housing Needs* (Alvarez & Steffen, 2021; Belsky et al., 2005). This measure captures households that meet four criteria:

a   earn 50% or less of the area median income,
b   spend more than 50% of their income toward housing,
c   do not receive housing assistance, and
d   live in crowded or physically inadequate housing.

In 2019, 7.8 million households had worst case housing needs according to a report issued to the U.S. Congress (Alvarez & Steffen, 2021). These households face acute housing challenges that are currently not met with existing housing support programs. The fact that more than 6% of all U.S. households face worst case housing needs emphasizes the significant affordable housing crisis in this country.

Location, Location, Location

In 2019, the median cost of a two-bedroom apartment in San Francisco was $2,314, while in Detroit, the same sized apartment rented for only $840 (2019 American Community Survey 1-Year Estimates). In addition to the stark differences in housing markets, we also observe significant wage differences between each city. Median household income in San Francisco during the same year was $123,859; in Detroit it was $33,965. As a result, the intuitive assumption that housing cost burdens are much higher in San Francisco than in Detroit proves to be incorrect. 2019 cost burdens were actually *higher* in Detroit than in San Francisco (excluding utility costs, housing cost burdens based on median income and rents were 22% in San Francisco and 30% in Detroit). While housing in Detroit is far cheaper than in San Francisco, wages are even lower on a relative basis. When discussing housing affordability, both Detroit and San Francisco have affordability problems, but the nature of those problems is different. Detroit has an income problem – wages are too low – while San Francisco has a cost problem – rents are too high.

This comparison highlights the multiple pathways to housing affordability problems. Importantly, the different pathways also underscore the need for solutions and responses that are sensitive to local context. The path to ensuring housing affordability in Detroit may look very different than the approach used in San Francisco. While complicated, solutions to the housing affordability crisis must acknowledge the different contexts in which these housing affordability challenges occur. As depicted in the comparison of San Francisco and Detroit, the analysis of housing cost burden is complicated because both housing costs and income must be considered. To understand and appreciate the dynamics of housing affordability, one must study both the housing market *and* the labor market (where wages are established). Further complicating this effort is the fact that labor markets and housing markets are highly localized.

Let's return to our comparison of Detroit and San Francisco. Imagine that you are a recent graduate with a master's degree in education, and you hope to build a career as an elementary school teacher. Due to your personal preferences and social networks, you have targeted Detroit and San Francisco as potential places to live and work. Because of the relatively modest earnings of early-career public school teachers, understanding the cost of living in each location is an important input into your decision-making process. After conducting some basic due diligence from the National Education Association website, you learn that an average teacher in Michigan earns $64,000. The similar figure in California is approximately $86,000. You are now faced with an interesting decision. If you based your decision solely on financial considerations, the

Table 1.1 Income thresholds for LIHTC program for Chicago MSA

Income Limit	Size of Household (# of people)							
	1	2	3	4	5	6	7	8
80% AMI	$52,240	$59,680	$67,120	$74,560	$80,560	$86,560	$92,480	$98,480
60% AMI	$39,180	$44,760	$50,340	$65,240	$70,490	$64,920	$69,360	$73,860
50% AMI	$32,650	$37,300	$41,950	$46,600	$50,350	$54,100	$57,800	$61,550
30% AMI	$19,590	$22,380	$25,170	$27,960	$30,210	$32,460	$34,680	$36,930

(*Source*: HUD 2021)

evidence presents mixed signals. You could earn a higher income in San Francisco, but your purchasing power (and potentially your standard of living) would be greater in Detroit. Certainly, there are a range of factors (beyond income and housing costs) on which one might rely when selecting a place to live, but this relatively simple example highlights how the interaction between housing and labor markets might prompt different decisions for different households.

One way to normalize or standardize housing affordability in different geographic contexts is the use of relative (and local) measures of income. A frequently used concept is known as Area Median Income (AMI). Categories have been created using the AMI as a benchmark (Aurand et al., 2021). For example, Extremely Low Income (ELI) households are those with incomes below the federal Poverty Guideline or 30% of AMI, whichever is greater. Very Low income (VLI) households have incomes above the ELI threshold and below 50% of AMI. Low Income (LI) households have incomes between 51 and 80% of AMI. Middle Income (MI) households are between 81 and 100 of AMI, while the remainder of households have incomes greater than the median. By definition, half of all households will have incomes above the AMI. Therefore, when affordability is discussed, one can discuss the intended targets of a program or policy based on the level of AMI that it is designed to address. For example, the Low Income Housing Tax Credit (LIHTC) program is the primary federal housing production program. Housing constructed with tax credits must maintain "affordability" for a period of at least 30 years, and a developer may use tax credits to construct housing to serve households with incomes up to 80% of AMI so long as the average income of all households served by the project is below 60% of AMI. Because households are different sizes, these income thresholds are based on household size.

HUD uses these concepts to establish income eligibility for public programs. Table 1.1 was used to determine income thresholds in 2021 for the Low Income Housing Tax Credit Program in the Chicago-Joliet-Naperville, IL Metropolitan Statistical Area (MSA). For housing to be affordable, a household should not pay more than 30% of household income toward housing in any month. Using the Chicago MSA as an example, a Low Income (LI) one-person household with household income at 50% of AMI could only spend $816 per month on housing and utilities to be considered affordable.[4]

Achieving Affordability

A final question that warrants attention in this introductory chapter is how does housing become affordable? If society desires to provide a greater number of affordable housing options to more households, how can it achieve that outcome? In some cases, housing is affordable because of naturally occurring factors. One such factor is low levels of rent. This housing may be in a low-cost neighborhood or in a city where housing costs are modest relative to wages. Another naturally occurring pathway to affordability is based on housing quality. As housing becomes older and its quality diminishes, its cost will fall. Outside of these "natural" processes in the private market, there are methods used by the government and non-profit organizations to provide or ensure affordability. First, some housing may be constructed with the explicit purpose of providing affordability to tenants and owners. This housing is frequently constructed directly by the government (e.g., public housing) or with the help of public subsidies or tax credits that allow developers and owners to deliver housing that is affordable to households with below average incomes. In addition to supply-side sources of affordable housing (supports that help increase the supply of housing), there are demand side solutions as well. Demand side supports provide purchasing power to recipients. Most commonly, governments may provide households with rental subsidies to allow them to consume housing that they otherwise would not be able to afford.

Finally, governments may also use regulatory tools to promote or ensure affordability. Examples include rent control or other tenant protections that limit rents that can be charged in the private housing market. Governments also have a range of rules and regulations related to zoning and permitting that can either inhibit or promote the construction of housing. Tight housing regulations can slow the construction of housing which can serve to increase its cost given constrained supply. Relaxing these regulations can, therefore, be used to increase access to housing and to lower its cost (or to moderate increases in housing costs). Local jurisdictions may also mandate inclusionary zoning which requires developers to provide a small percentage of units in each new building that are affordable for lower AMI households. There is a robust debate about the efficacy of inclusionary zoning as a tool to provide affordable housing, but it is an important regulatory tool in nationwide efforts to increase the stock of affordable housing.

When discussing affordable housing, much of the focus and attention is directed toward the rental market. This focus makes intuitive sense; many households with incomes below 80% of AMI frequently rent their housing. Homeownership is more prevalent among middle- and higher-income households in the U.S. But it is important to highlight that homeownership may also provide a pathway to consistent, affordable housing and as a tool for wealth creation; once a household owns its house, they are immune to the pressures of rising prices (but they are potentially exposed to rising property taxes which could jeopardize affordability). In this way, homeownership provides greater protection, and more security, than year-to-year rental leases that expose households to the vagaries of the rental housing market. One only need to look at the rapid increase in rental costs after the Covid-19 pandemic to see the market risk to which renter households are exposed. This is one reason why affordable homeownership is an increasing area of focus for many housing advocates.

Book Synopsis

As this introduction demonstrates, there are many complexities and nuances to be explored within the broader concept of affordable housing. To provide readers with a deeper understanding of affordable housing in the United States, this book is broken out into five sections. In the second section, Chapters 2 through 6, we provide an overview of the housing system in the United States and affordability of housing across the country. Chapter 4 covers the history of affordable housing in America as a base from which to understand the existing regulations, programs, and policies that shape current approaches to housing delivery and support. The last two chapters in this section address the implications and consequences of a lack of affordable housing. Chapter 5 considers issues of race and inequality in housing, particularly affordable housing, and the adverse effects of unaffordable housing on individual and household well-being. In the final chapter of this section, we summarize negative housing outcomes related to a lack of affordable housing, including foreclosure, eviction, and homelessness.

In the third section, we analyze the various ways in which entities provide access (albeit at an insufficient scale) to affordable housing either through public or private sector action. In Chapter 7, we describe the various actors involved in the affordable housing system – including local and federal governments, regulatory agencies, non-profit organizations, and for-profit developers – and consider how each of these actors interact to promote and deliver affordable housing. In Chapter 8, we summarize supply-side strategies such as public housing, tax credits, and other production subsidies. We then describe demand-side strategies including income supplements and housing vouchers in Chapter 9. In Chapter 10, we introduce affordable homeownership and the various models that provide ownership opportunities for lower income households. We conclude this section by describing the important role that regulations play in the affordable housing landscape.

In the fourth section, we provide three case studies that summarize how the complex web of organizations, policies, and market conditions interact to produce affordable housing or the lack thereof. By focusing on Chicago, IL, Seattle, WA, and San Antonio, TX, we highlight the substantial regional variation in housing affordability and how the pathways and policy responses differ by location.

We conclude the book in its fifth and final section by charting a path forward. In the closing chapter, we propose a roadmap for housing affordability at the national, state, and local level, while recognizing the substantial regional variation that exists. The large scale of the affordable housing crisis in the United States requires responses of a similar magnitude. Modest or incomplete solutions may help a select few, but a comprehensive response that would meet the significant need requires bold action, commitment, and resources from all sectors of society.

Notes

1. The following books are excellent complements to this volume: Schwartz (2021), Scheutz (2022), Bull & Gross (2023), and Anacker et al. (2018).
2. Extremely Low Income (ELI): households with income at or below the federal poverty guideline or 30% of Area Median Income (AMI), whichever is greater.
3. For example, the Biden Administration announced new actions to address housing affordability challenges. www.whitehouse.gov/briefing-room/statements-releases/2024/02/29/fact-sheet-biden-harris-administration-announces-new-actions-to-boost-housing-supply-and-lower-housing-costs/
4. $32,650 annual income equals $2,721 monthly income. 30 percent of monthly income is $816.

References

Airgood-Obrycki, W., Hermann, A., & Wedeen, S. (2022). "The rent eats first": Rental housing unaffordability in the United States. *Housing Policy Debate*, *33*(6), 1272–1292.

Alvarez, T., & Steffen, B. L. (2021). *Worst case housing needs 2021 report to congress.* Washington, DC: U.S. Department of Housing and Urban Development. Retrieved from www.huduser.gov/portal/publications/Worst-Case-Housing-Needs-2021.html

Anacker, K. B., Carswell, A. T., Kirby, S. D., & Tremblay, T. (2018). *Introduction to housing*, 2nd edition. The University of Georgia Press.

Aratani, Y., Chau, M., Wight, V. R., & Addy, S. (2011). *Rent burden, housing subsidies and the well-being of children and youth*. New York, NY: National Center for Children in Poverty, Columbia University.

Australian Housing and Urban Research Institute (AHURi). (2019). *Understanding the 30:40 indicator of housing affordability stress*. Retrieved from www.ahuri.edu.au/analysis/brief/understanding-3040-indicator-housing-affordability-stress#:~:text=What%20is%20the%2030%3A40,its%20income%20in%20housing%20costs.

Aurand, A., Emmanuel, D., Threet, D., & Rafi, I. (2021). *The gap. A shortage of affordable homes*. National Low Income Housing Coalition. Retrieved from https://reports.nlihc.org/sites/default/files/gap/Gap-Report_2021.pdf

Aurand, A., Pish, M., Rafi, I., & Yentel, D. (2022). *Out of reach. The high cost of housing*. National Low Income Housing Coalition. Retrieved from https://nlihc.org/sites/default/files/oor/2022/OOR_2022_Mini-Book.pdf

Belsky, E. S., Goodman, J., & Drew, R. (2005). *Measuring the nation's rental housing affordability problems*. Cambridge, MA: Joint Centre for Housing Studies, Harvard University. Retrieved from www.jchs.harvard.edu/research/publications

Bentley, R., Baker, E., & Aitken, Z. (2019). The 'double precarity' of employment insecurity and unaffordable housing and its impact on mental health. *Social Science & Medicine, 225*, 9–16.

Bull, M., & Gross, A. (2023). *Housing in America: An introduction*. UK: Routledge.

Colburn, G, Hess, C., Allen, R., & Crowder, K. (2024). The dynamics of housing cost burden among renters in the United States. *Journal of Urban Affairs*.

Colburn, G., & Aldern, C. P. (2022). *Homelessness is a housing problem: How structural factors explain U.S. patterns*. Oakland, CA: University of California Press.

Elliott, S., West, S., & Castro, A. B. (2021). Rent burden and depression among mothers: An analysis of primary caregiver outcomes. *Journal of Policy Practice & Research, 2*(4), 285–300.

Graetz, N., Gershenson, C., Porter, S. R., Sandler, D. H., Lemmerman, E., & Desmond, M. (2024). The impacts of rent burden and eviction on mortality in the United States, 2000–2019. *Social Science & Medicine, 340*, 116398.

Ezennia, I. S., & Hoskara, S. O. (2019). Methodological weaknesses in the measurement approaches and concept of housing affordability used in housing research: A qualitative study. *PloS ONE, 14*, e0221246.

Herbert, C., Hermann, A., & McCue, D. (2018). *Measuring housing affordability: Assessing the 30 percent of income standard*. Cambridge, MA: Joint Center for Housing Studies of Harvard University.

Kingsella, M., & MacArthur, L. (2022). *Housing underproduction in the U.S.* Retrieved from https://upforgrowth.org/wp-content/uploads/2022/09/Up-for-Growth-2022-Housing-Underproduction-in-the-U.S.pdf

Leishman, C., & Rowley, S. (2012). "Affordable housing." In D. F. Clapham, W. A. V. Clark, & K. Gibb (Eds), *The Sage handbook of housing studies*. Thousand Oaks. CA: Sage Publishing.

National Low Income Housing Coalition. (2024). *The gap: A shortage of affordable homes*. Retrieved from https://nlihc.org/gap

Schanzenbach, D. W., Nunn, R., Bauer, L., & Mumford, M. (2016). *Where does all the money go: Shifts in household spending over the past 30 years*. Washington, DC: The Hamilton Project, Brookings.

Schuetz, J. (2022). *Fixer-Upper. How to Repair America's Broken Housing System*. Washington, DC: Brookings Institution Press.

Schwartz, A. (2021). *Housing policy in the United States*, 4th edition. Routledge.

Solomon, D., Maxwell, C., & Castro, A. (2019). *Systematic inequality: Displacement, exclusion, and segregation. How america's housing system undermines wealth building in communities of color*. Washington, DC: Center for American Progress.

Stone, M. E. (1993). *Shelter poverty: New ideas on housing affordability*. Philadelphia: Temple University Press.

Stone, M. E. (2006). What is housing affordability? The case for the residual income approach. *Housing Policy Debate, 17*(1), 151–184.

Stone, M. E. (2012). "Shelter Poverty." In A. T. Carswell (Ed.), *The encyclopedia of housing*, 2nd edition. Thousand Oaks, CA: Sage Publishing.

U.S. Housing and Urban Development (HUD). (2014). *Rental burdens: Rethinking affordability measures*. Retrieved from www.huduser.gov/portal/pdredge/pdr_edge_featd_article_092214.html

U.S. Housing and Urban Development (HUD). (2017). *Defining housing affordability*. Retrieved from www.huduser.gov/portal/pdredge/pdr-edge-featd-article-081417.html#:~:text=Keeping%20housing%20costs%20below%2030,to%20be%20housing%20cost%20burdened.

U.S. Housing and Urban Development (HUD). (2021). *FY 2021 multifamily tax subsidy project income limits summary*. Retrieved from www.huduser.gov/portal/datasets/il/il2021/2021sum_mtsp.odn?inputname=METRO16980M16980*Chicago-Joliet-Naperville%2C+IL+HUD+Metro+FMR+Area&area_choice=hmfa&year=2021

Zonta, M. (2020). *Expanding the supply of affordable housing for low-wage workers*. Center for American Progress. Retrieved from www.americanprogress.org/article/expanding-supply-affordable-housing-low-wage-workers/

Part II

Understanding Affordable Housing

Chapter 2

The Housing Stock in the United States

Overview

According to the 2020 Census, there are over 140 million housing units in the United States. Given a population of 331 million, this suggests an average of 2.4 people in each housing unit across the nation. Of course, averages mask significant variation. Many people live alone, some reside in crowded households, and according to the 2023 point-in-time homeless count, over 650,000 people experienced homelessness. The housing experiences of people living in the United States are as diverse as the nation itself. Providing national statistics about housing is important context for the study of affordable housing, but there is not just one market for housing in the United States. Markets for housing and real estate are local. To understand housing in the U.S., therefore, one must not focus solely on national averages and trends, but rather understand how income and housing access interact in the specific locations where people live.

Housing Tenure and Types

The housing stock in the United States is dominated by single-family, detached homes. Over 62% of all housing units are detached and are designed to house one family (2021 ACS 1-Year Estimates). As discussed later in this book, the importance of single-family housing in the development of the United States cannot be understated. The dream of a detached home with a white picket fence is embedded in the culture of the nation as well as its housing policy. The disproportionate percentage of detached, single-family homes in the U.S. is not an accident of the private housing market. Rather, the housing system was structured to produce this outcome. The remaining 38% of housing units are town homes, multi-family dwellings of different sizes, or less common housing types like manufactured homes or house boats. Only a relatively small percentage of housing units – just over 10% – are part of large buildings that have over 20 units in the building. The lack of housing density is another prominent feature of the U.S. housing system, especially when compared to other countries.

We can also break down the housing stock by different types of tenure arrangements. Housing tenure is the legal arrangement by which a household secures the right to occupy their housing unit. Owner occupancy is the dominant form of tenure in the United States; 65% of all housing units are owned (2021 ACS 1-Year Estimates). The remaining 35% of units are rented, and the vast majority of those units are rented in the private market. Figure 2.1 depicts the homeownership rate in the United States over the last several decades (U.S. Census Bureau, 2023). Over time, roughly two thirds of all housing units have been owned in the nation. The percentage approached 70% in 2004 in the midst of the housing boom that preceded the foreclosure crisis and Great Recession of 2007 to 2009, and the level of ownership fell to below 64% in 2016.

DOI: 10.1201/9781003356585-4

Figure 2.1 U.S. Homeownership Rates from 1964 to 2021
(Source: U.S. Census Bureau, 2023)

The U.S. is marked by stark differences between ownership and renting, and these differences are not solely contractual. Social science research highlights preferences for ownership as compared to renting housing and other goods (Kricheli-Katz & Posner, 2023). Goetz and Sidney (1994) noted the "ideology of property" in the United States that privileges ownership at the expense of renters. In an article entitled, "The grapes of rent: A history of renting in a country of owners," Krueckeberg (1999) summarized the disconnect between owners and renters in the U.S.:

> I argue that from colonial times to the present, a bias in favor of property ownership has been prominent in American public policy. Over time, this bias has manifested itself in several ways: property requirements that denied renters the rights of suffrage in colonial times, land distribution schemes that aggrandized ownership rather than settlement, and finally a variety of regressive federal and state tax policies in which homeowners and the real estate fraternity are enormously subsidized by tenants who are excluded from the largesse they help supply.
> (p. 11)

Thus, in the U.S., there is a long history of stigma associated with rental housing while homeownership has become the socially desirable form of housing tenure for many households.

In Table 2.1 below, we break down the housing stock by tenure (ownership versus rental) and housing type. This table highlights some notable differences. First, homeownership occurs disproportionately in detached, single-family homes. Almost 83% of owned housing structures are of this variety. Renting is distributed far more evenly over a range of housing types, from single-family homes to units in large apartment buildings.

There are significant cleavages in the housing market based on ownership status, but gaps also emerge when the housing stock is broken down by the race and income of inhabitants. Housing type and tenure are not equally distributed throughout society. Extensive scholarship has highlighted the significant role of racism and discrimination in the housing market. People

Table 2.1 U.S. Housing Stock in 2021

	Single-family		Multi-family			Other
	Detached	Attached	2–4 units	5–19 units	20 or more units	RV, boat, van, mobile home, etc.
Ownership	82.5%	6.3%	2.0%	1.4%	2.0%	5.7%
Rental	25.8%	6.6%	17.4%	22.1%	23.9%	4.2%
Total	62.9%	6.4%	7.4%	8.5%	9.6%	5.2%

(*Source*: American Community Survey (ACS), 2021)

of color have faced many barriers in the housing market over time: the Federal Housing Administration's redlining practices restricted access to certain neighborhoods (Rothstein, 2017); the housing finance system made it difficult for households of color to secure a mortgage, thus dramatically limiting homeownership opportunities (Krivo & Kaufman, 2004; Kuebler & Rugh, 2013); and the siting of public housing projects and affordable housing in primarily low-income neighborhoods occupied by households of color concentrated race and economic disadvantage in urban ghettos (Rohe & Freeman, 2001). These factors, as they played out over history, have produced very different housing outcomes for households with different demographic characteristics. Table 2.2 disaggregates Table 2.1 by race. When we compare white and Black households, conspicuous differences emerge. White households are far more likely to own their homes, and are more likely to live in detached, single-family homes. Black households reside disproportionately in multi-family housing, and do so primarily as renters. These differences have profound consequences in terms of wealth creation, as well as access to quality schools, employment, and transportation. In sum, the U.S. housing market can be split into two categories: Homeownership, which disproportionately serves higher income, white households, and the rental market that most frequently serves households of color and those with lower incomes. We devote Chapter 5 to the discussion of racism and discrimination in the housing system, but these differences are important to highlight when discussing the housing stock in the U.S.

We continue the analysis by breaking down housing tenure by income. As Table 2.3 below demonstrates, there is a strong relationship between income and homeownership. Rates

Table 2.2 U.S. Housing Stock in 2021, By Race

	Tenure		Single-family		Multi-family			Other
	Owner	Renter	Detached	Attached	2 to 4 units	5 to 19 units	20 or more units	RV, mobile home, boat, van, etc.
White	73.3%	26.7%	69.2%	5.7%	5.6%	6.4%	7.6%	5.5%
Black	44.0%	56.0%	46.7%	8.6%	12.0%	15.3%	13.9%	3.5%
Asian*	62.2%	37.8%	53.3%	10.1%	7.6%	10.6%	17.4%	1.1%
Hispanic	50.6%	49.4%	52.6%	6.6%	10.9%	11.5%	12.4%	6.1%
Other race	52.2%	47.8%	53.7%	6.4%	10.7%	11.1%	11.7%	6.3%
Total	63.7%	36.3%	61.7%	6.4%	7.8%	8.9%	9.9%	5.3%

(*Source*: ACS, 2021)

* Includes Pacific Islander

Table 2.3 U.S. Housing Tenure in 2021, by Income

Household income	Owner occupied	Renter occupied
less than $20,000	5.5%	8.0%
$20,000 to $50,000	12.2%	10.8%
$50,000 to $150,000	32.5%	13.4%
$150,000 or more	15.2%	2.5%
Total	65.4%	34.6%

(Source: ACS, 2021)

of homeownership are far higher for households with higher levels of income. And given that home equity is a primary source of household wealth generation in the U.S., we can see how homeownership exacerbates wealth inequality. Higher income households are able to purchase homes and are, therefore, able to benefit from the wealth creation derived from the increase in home values.

An open question is whether the U.S. has the right mix of housing for the way people now live. People are partnering at lower rates, and when they do, they do so later in life. Recent data from U.S. Census demonstrates that nearly 28% of all households comprise a single person, which is a record high (Anderson et al., 2023). This change alone suggests a greater need for diverse housing options, including apartment-style living for the millions of people who live alone for longer periods in adulthood. Notions of the American dream have also shifted. While homeownership remains a goal for a majority of Americans, affordability is an increasingly significant impediment to achieving that dream (National Association of Realtors, 2017). Surveys also show that there is a desire for a greater diversity of housing types throughout the lifecycle that is currently not met by the existing housing stock. Many households prefer smaller attached units in denser locations near amenities and at more affordable prices, even though the dominant form of new housing built remains single-family homes (Logan & Mangold, 2018). The rapid increase in multi-family development over the last decade highlights how the residential built environment is responding to changing needs and preferences.

To conclude, we place the U.S. in a brief international comparison. There are both similarities and differences between the U.S. and other developed economies. In terms of tenure, the United States is fairly average. Several eastern European nations (former Soviet bloc countries) have homeownership rates in excess of 90% which was the result of formerly state-owned housing units that were transferred to private ownership following the collapse of the Soviet Union. On the low end, Germany and Switzerland have some of the lowest homeownership rates among developed countries, as rental housing is a more dominant and socially acceptable form of tenure in those nations (Goodman & Mayer, 2018). The U.S. is in the middle for homeownership, but when we compare the type of housing structures that are dominant in a country, the U.S. stands out materially. Global comparisons highlight that the level of single-family detached housing and lack of density make the U.S. highly unique. For example, detached dwelling units are 52% of all housing units in Japan and only 28% in Germany (OECD Affordable Housing Database). In the U.S., the comparable figure is nearly 63% (2021 ACS 1-Year Estimates). This lack of density and multi-family housing presents challenges when it comes to providing and ensuring affordable housing for all who need it in the U.S.

Housing Quality in the United States

We next explore the age and quality of the U.S. housing stock. Despite the rapid population growth of the United States over the last two to three decades, 75.8% of all housing units were constructed before the year 2000 (2021 ACS 1-Year Estimates). Over half of all housing was constructed before 1980, which, given the strong focus on single-family homes as a key element of the American Dream during the middle of the 20th century, helps to explain the disproportionate share of detached, single-family homes in the U.S. housing stock. For owner-occupied housing – which is predominantly detached, single-family – the average age of housing in 2021 was 40 years (2021 ACS 1-Year Estimates). Because housing construction has failed to keep up with population growth, the average age of housing continues to rise. In 2005, the average age of owner-occupied housing was only 31 years. So, what is the typical lifespan of the existing housing stock in the U.S.?

Housing is a durable asset; historically, many homes have had a useful life of over 100 years. Existing research based on U.S. data suggests an average lifespan between 60 (Aktas & Bilec, 2012) and 130 years (Ianchenko et al., 2020). While long, this estimate is shorter than housing in other countries due to different building standards in the U.S. Also, just because a house may last 100 years does not mean that all systems in the house will be equally durable. Plumbing, roofing, heating, siding, and electrical systems frequently become obsolete while a home is still operational. Replacing these systems adds to the cost of owning – or living in – an older home.

The general durability of housing raises an interesting conundrum. Decisions we make about our housing stock today will have a strong influence on the housing system for the next 100 years. Given rapidly changing demographics, household structures, and preferences, it is a challenge to construct the right mix of housing that will be occupied decades in the future by people who are not even born at the time of development. Many suburban neighborhoods are filled with post-World War II single-family housing that no longer conforms to modern design preferences and aesthetics. Will the housing constructed today be of interest to renters and home buyers in 2070?

Despite the age of the U.S. housing stock, its quality is quite high. In the years prior to World War II, housing quality was a major social problem. Millions of people resided in housing with inadequate plumbing, heating, and insulation (von Hoffman, 2012). Unlike today, the primary housing issue for low-income Americans was the issue of quality, *not* affordability. Over the succeeding decades, new building regulations and codes gradually helped to improve the overall quality of housing in the United States. In 1920, only 1% of housing units in the U.S. had electricity and indoor plumbing. Following the implementation of plumbing codes, the number of housing units without adequate plumbing fell to just a few percent by 2000 (Lutz, 2004). According to data from the American Housing Survey, in 2019, 3.6% of all housing units were deemed moderately inadequate, while only 1.2% of units were severely inadequate. Therefore, despite the fact that a majority of people reside in housing that was built before Ronald Reagan became President, the quality of that housing is generally satisfactory. Recent research, though, raises the question of whether the American Housing Survey appropriately captures the concept of housing adequacy; critics suggest that the survey may underestimate the amount of inadequate housing in the United States (Newman & Garboden, 2013).

As housing quality has risen over time, so has the size of housing that most households occupy. In general, newer housing is larger than the housing that was built in prior generations. Detached single-family homes built in the 1960s averaged 1,500 square feet while the same type of housing constructed in the 2000s averaged about 2,200 square feet – a nearly 50% increase, while the average number of rooms also increased. Multi-family housing units increased in size over time, from 800 square feet in the 1960s to 1,200 square feet by 2009, even as the average number of rooms (four) stayed constant (Sarkar, 2011). Americans generally live in larger and less crowded living conditions than they did in prior generations.

Housing Supply and Affordability

This chapter has described the housing stock in the United States. We have broken down the housing stock by housing type and tenure and also discussed its age and quality. What we have not done is address the overall adequacy of this stock. Do we have enough housing? The data suggest the answer is no. Recent analysis from Freddie Mac concludes that the 141.2 million housing units in 2020 were 3.8 million units short of the nation's need (Khater, 2021). A particularly troubling statistic from this study is that the supply deficit grew by over one million units between 2017 and 2020. Failure to close this gap could have significant consequences for society in the years and decades to come. In particular, Khater (2021) highlighted the significant deficit of smaller, starter homes for the large cohort of Millennials that are now approaching homebuying age. Without sufficient supply, demand drives up prices and has negative implications for affordability.

Figure 2.2 below traces housing construction in the United States over the last five decades. Even as the nation's population has grown, the level of housing construction has waned. This housing shortage is a troubling national trend that accelerated over the last two decades. At the onset of the Great Recession, housing construction fell dramatically. Tighter

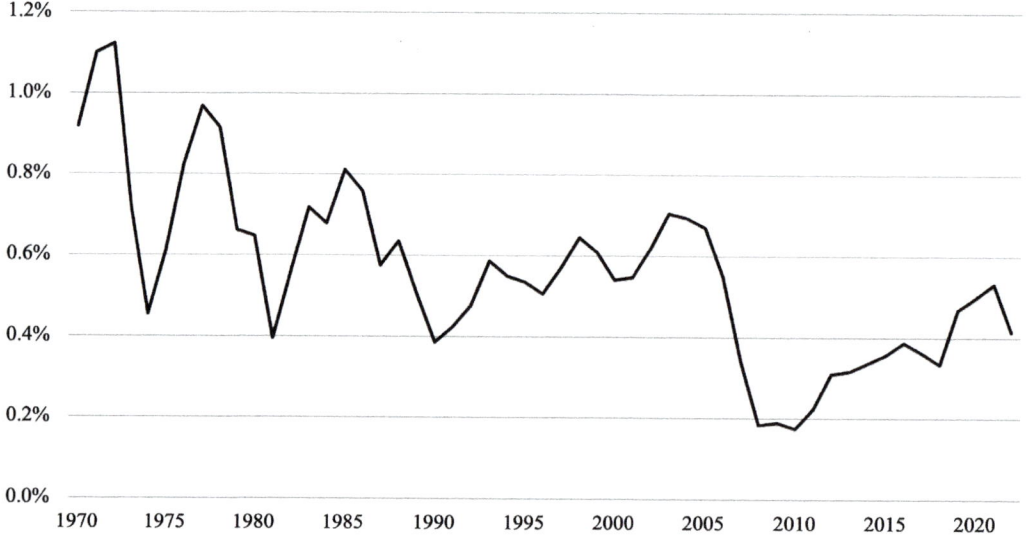

Figure 2.2 U.S. Housing Starts as a Share of the Population
(Source: U.S. Census Bureau)

mortgage financing markets and challenging economic conditions were to blame, and this decade-long drop in housing production has contributed to shortages that we now observe in the 2020 data.

Affordability has now taken center stage as the primary housing concern in the United States. Early in the 20th century, concerns over housing quality were at the forefront, but those issues have largely been addressed through building codes and improved construction practices. Over the last 100 years, the United States has mandated higher-quality housing – but has done so without ensuring that the entire population can afford housing constructed and maintained at that level. When combined with the chronic supply shortages, a significant affordability crisis has emerged.

In the following chapter, we take a deeper dive into housing affordability dynamics in the United States. To conclude this section, we provide an introduction to the scale of the affordable housing crisis in the United States. Using data from the National Low Income Housing Coalition (NLIHC), we demonstrate that a lack of affordable housing is one of the most pressing issues facing the nation.

The shortage of housing in the U.S. is far more pronounced for households with lower incomes. According to research from the NLIHC, there is a shortage of 7.3 million housing units for extremely low-income renters – households that earn less than 30% of the area median income – throughout the U.S. (Aurand et al., 2023). Only about a third of extremely low-income households have access to housing that is affordable and available. The ongoing failure to address this acute need places low-income households in a difficult position. Without housing support from the government, the options available to these households are limited and undesirable. They can either pay a disproportionate share of their limited income toward housing, leaving less to spend on other household necessities, or they can move into someone else's home – a concept referred to as doubling up – to attain affordability by crowding into one housing unit. Neither option is tenable and contributes to the millions of households that are precariously-housed in the U.S. – a number far higher than the annual point-in-time estimates of homelessness published annually by HUD.

A final point that we have not addressed in this chapter, but that is central to this discussion, is who constructs and operates the housing stock in the U.S. As described in the previous paragraph, for low-income households seeking housing in the private housing market, options are limited. But are there other options besides housing constructed and operated by private developers and landlords? The answer is yes, but at a limited scale. In all countries, the public sector plays an important role in the housing system. Governments in the U.S. provide housing support through multiple mechanisms:

1 they construct publicly owned housing for residents, frequently called public or social housing;
2 they provide subsidies to support the construction of affordable housing for low-income households; and
3 they provide rental subsidies to allow low-income households to procure housing in the private market.

The primary federal housing programs in the U.S. – Public Housing, Low-Income Housing Tax Credits, and Housing Choice Vouchers – provide subsidized housing units and rental subsidies to 970,000, 3.5 million, and 2.2 million households, respectively.[1] While at first glance these may seem like significant numbers, the share of federally subsidized households in the U.S. is only approximately 5%.[2] This compares to a national poverty rate of roughly 12%. Among households that are *eligible* for housing support from the federal government, only about one in four receive assistance (HUD, 2021). This statistic highlights the limited nature of housing supports in the United States. The remaining households are left to fend for themselves in the private housing market. The limited scale of federal housing supports leaves millions of U.S. households with demonstrable needs fighting to secure housing in the private rental market. By comparison, the stock of housing that is publicly-owned or subsidized is between 10 and 50% in many Northern and Western European countries (Milligan & Gilmour, 2012). In sum, the U.S. and other Anglo countries (e.g., Canada and Australia) are outliers relative to other advanced economies given the limited involvement of the public sector in the nation's housing system. And this policy decision has had profound consequences for low-income households throughout the United States.

Beyond the federal assistance described in the preceding paragraph, there are also state and local housing assistance programs that support low-income households. States and cities offer programs that provide direct rental assistance and capital to expand the rental housing stock. It is difficult to capture the extent of these programs since they are administered at the local level and there is no reporting mechanism that quantifies the impact of these patchwork measures. However, every few years, the National Low Income Housing Coalition attempts to identify all active programs to populate their publicly available State & City Funded Rental Housing Programs database. There are currently over 150 programs included in the database. The U.S. Census Bureau in the Annual Survey of State and Local Government Finances identifies state and local expenditures on housing and community development. Around 2% of direct expenditures is spent on housing and community development, which is approximately $59 billion (Urban Institute, 2020). Most of these expenditures are dedicated to rental assistance and operational costs for housing programs and only a small portion is used for the production of affordable housing. While helpful, one of the limitations of this local support is its scale. Subsidies provided by state and local sources are shallow, as opposed to the relatively deep subsidies provided by the federal government.

Housing Construction

Thus far in this chapter we have described the housing stock of the U.S. and its various attributes, but we have not considered how this supply of housing has been developed and constructed. Generally speaking, we construct housing in the 2020s very similarly to the way we did in the 1960s. There has been a glaring lack of technological advancement in the construction industry compared to other fields like technology and manufacturing (Potter, 2021). A consequence of this lack of innovation is that prices to construct housing have not fallen as they have for many other goods. For example, the cost to manufacture a television or computer has fallen significantly (on an inflation-adjusted basis) over the last half century even as quality and functionality have soared. Our food production has also become more efficient as yields (production per acre) have increased dramatically (Ritchie et al., 2021). Housing, therefore, is an industrial outlier; the industry's failure to innovate and adapt has contributed to the high cost of new housing development and ongoing deficits in housing supply (Ivory & Colton, 2020). Even so-called affordable housing is incredibly expensive to construct. Constructing housing affordable to those with below median incomes using the LIHTC program can be costly for developers, primarily due to financing complexity and regulatory requirements. For projects placed in service from 2011–2016, the median LIHTC development cost per-unit was $164,757 (Lubell & Wolff, 2018). And in expensive markets such as the San Francisco Bay Area, development costs can exceed $600,000 per unit (Reid et al., 2020). Any housing is expensive to construct but there are multiple culprits for the additional costs to construct affordable units, including the administrative burden associated with the LIHTC program[3] and other funding sources, and local regulations that govern housing construction (Kneebone & Reid, 2021; Reid et al., 2020). Reforms and innovation that could lower the price to construct new housing are essential if we are going to address the substantial shortage of housing in the United States. We explore housing development and the regulations that govern that activity in greater detail in Chapter 11.

A common question that arises when discussing housing construction is what type of housing should be constructed. We have seen that most new residential construction is tailored toward higher income households. Large, single-family dwellings are built in the suburbs to meet the needs of affluent buyers, while in downtown neighborhoods, expensive condominiums and apartments are built to serve the highly compensated urban workforce. But does this development meet the need for more affordable housing? The answer is complicated, but, in the short-run, this housing may have a modest impact on affordability. There is a frequently cited concept in the field of planning and housing called "filtering." The idea behind filtering is that the housing market works as a single system; housing constructed at the high end of the market eventually ages and becomes more affordable. Under this theory, as more market rate units get built, previously built units will eventually filter down to create the affordable housing of the future. While intuitive, the empirical support for filtering is scant (Zuk & Chapple, 2016). At best, filtering is a generational response to housing affordability and is not an effective response to the affordable housing crisis at hand. Research demonstrates that market rate development alleviates some pressure on the housing market. When expensive units are in short-supply, high income households will consume less expensive housing than they otherwise would, thus limiting the supply of housing for moderate and low-income households. This finding highlights that the shortage of housing, at all income levels, is a problem that can affect housing affordability. Therefore, a robust response to the affordable housing crisis in the United States is to construct housing that is accessible to households at all income levels. Comprehensive strategies are needed to address gaps in supply within each housing submarket. Failure to do so will likely exacerbate current shortages and expand the affordability crisis in the United States.

Conclusion

As documented in this chapter, the housing stock in the United States is inadequate, especially for households with lower levels of income. Given the limited support from the federal government, it is a challenge to increase the supply of affordable housing. In response, a few options exist:

1 local governments could preserve the supply of naturally-occurring affordable housing (NOAH) in the private rental market to serve low-income households;
2 private developers could build smaller and more cost-efficient housing that would be accessible to lower income households;
3 technological and process innovation could dramatically reduce the cost of construction and, hence, reduce the cost of rental housing for households that need it; and

4 the government could play a larger role in subsidizing either the construction of housing or lower income households in an effort to improve affordability.

In reality, some combination of all of these efforts will likely be needed to substantially improve the availability and affordability of housing in the United States.

Notes

1 We discuss these programs in detail in Chapters 8 and 9.
2 As described in Colburn et al. (2024), there is significant overlap between the Housing Choice Voucher and LIHTC programs, i.e., many people residing in LIHTC units also receive a housing voucher. As a result, the number of subsidized households in the U.S. is less than the sum of the total number of public housing units, LIHTC units, and vouchers.
3 Greater detail on the LIHTC program and its administrative burdens are described in Chapter 8.

References

Aktas, C. B., & Bilec, C. C. (2012). Impact of lifetime on US residential building LCA results. *The International Journal of Life Cycle Assessment, 17*(3), 337–349.

Anderson, L., Washington, C., Kreider, R. M., & Gryn, T. (2023). *Share of one-person households more than tripled from 1940 to 2020.* United States Census Bureau. Retrieved from www.census.gov/library/stories/2023/06/more-than-a-quarter-all-households-have-one-person.html

Aurand, A., Emmanuel, D., Foley, E., Clarke, M., Rafi, I., & Yentel, D. (2023). *The GAP: A shortage of affordable homes, March 2023.* The National Low Income Housing Coalition, Washington, DC.

Center on Budget and Policy Priorities. (2022). *Federal rental assistance fact sheets.* Retrieved from www.cbpp.org/research/housing/federal-rental-assistance-fact-sheets#US

Colburn, G., Acolin, A., & Walter, R. (2024). Subsidy overlaps in federal housing policy. *Housing Policy Debate.*

Freddie Mac. (2021). *Housing supply: A growing deficit.* Retrieved from www.freddiemac.com/research/insight/20210507-housing-supply

Goetz, E. G., & Sidney, M. (1994). Revenge of property owners: Community development and the politics of property. *Journal of Urban Affairs, 16*(4), 319–334.

Goodman, L. S., & Mayer, C. (2018). Homeownership and the American dream. *Journal of Economic Perspectives, 32*(1), 31–58.

Ianchenko, A., Simonen, K., & Barnes, C. (2020). Residential building lifespan and community turnover. *Journal of Architectural Engineering, 26*(3).

Ivory, A., & Colton, K. W. (2020). Innovative solutions for the housing crisis. *Stanford Social Innovation Review.* Retrieved from https://ssir.org/articles/entry/innovative_solutions_for_the_housing_crisis

Joint Center for Housing Studies. (2021). *The State of the Nation's Housing.* Cambridge, MA: Harvard University.

Khater, S. (2021). *One of the most important challenges our industry will face: The significant shortage of starter homes.* Washington, DC: Freddie Mac. Retrieved from www.freddiemac.com/perspectives/sam-khater/20210415-single-family-shortage

Kneebone, E., & Reid, C. (2021). *The complexity of financing low-income housing tax credit housing in the united states.* Terner Center for Housing Innovation, UC Berkeley.

Krivo, L. J., & Kaufman, R. L. (2004). Housing and wealth inequality: Racial-ethnic differences in home equity in the United States. *Demography, 41*, 585–605.

Krueckeberg, D. A. (1999). The grapes of rent: A history of renting in a country of owners. *Housing Policy Debate, 10*(1), 9–30.

Kuebler, M., & Rugh, J. S. (2013). New evidence on racial and ethnic disparities in homeownership in the United States from 2001 to 2010. *Social Science Research, 42*(5), 1357–1374.

Logan, G., & Mangold, K. (2018). *2018 housing and community preference survey.* RCLCO Real Estate Consulting. Retrieved from www.rclco.com/publication/2018-housing-and-community-preference-survey/

Lubell, J., & Wolff, S. (2018). *Variation in development costs for LIHTC projects.* Rockville, MD: Abt Associates.

Lutz, J. D. (2004). *Lest we forget, a short history of housing in the United States*. Ernest Orlando Lawrence Berkeley National Laboratory, LBNL-4751E.

Milligan, V., & Gilmour, T. (2012). "Affordable housing strategies." In S. J. Smith (Ed.), *International Encyclopedia of Housing and Home*. Amsterdam, Netherlands: Elsevier.

Napolitano, A., & Stambuk-Torres, B. (2020). *The costs of affordable housing production: Insights from California's 9% low-income housing tax credit program*. Terner Center for Housing Innovation, UC Berkeley.

National Association of Realtors. (2017). *2017 PULSE National Survey*. Retrieved from https://cdn.nar.realtor//sites/default/files/migration_files/reports/2017/national-pulse-report-2017–07–12.pdf?_gl=1*1t4rh6t*_gcl_au*NDc0MTA0MjgxLjE2OTU3MjcwMTU.

Newman, S. J., & Garboden, P. M. E. (2013). Psychometrics of housing quality measurement in the American Housing Survey. *Cityscape*, *15*(1), 293–306.

OECD Affordable Housing Database. (2021). *Residential stock by dwelling Type*. Retrieved from www.oecd.org/social/family/HM1–5-Housing-stock-by-dwelling-type.pdf

OECD Affordable Housing Database. (2021). *Housing conditions*. Retrieved from www.oecd.org/housing/data/affordable-housing-database/housing-conditions.htm

Potter, B. (2021). Why it's hard to innovate in construction. *Construction Physics*. Retrieved from https://constructionphysics.substack.com/p/why-its-hard-to-innovate-in-construction?s=r

Reid, C., Kricheli-Katz, T., & Posner, E. A. (2023). Ownership and rent stigma: Two experiments. *Behavioural Public Policy, 7*(2), 353–379.

Reid, C., Napolitano, A., & Stambuk-Torres, B. (2020). *The costs of affordable housing production: Insights from California's 9% low-income housing tax credit program*. Terner Center for Housing Innovation: UC Berkeley. Retrieved from www.ncsha.org/wp-content/uploads/2018/09/Final-LIHTC-Costs-Analysis_2018_08_31.pdf

Ritchie, H., Rosado, P., & Roser, M. (2021). *Crop yields*. Retrieved from https://ourworldindata.org/crop-yields

Rohe, W. M., & Freeman, L. (2001). Assisted housing and residential segregation: The role of race and ethnicity in the siting of assisted housing developments. *Journal of the American Planning Association, 67*(3), 279–292.

Rothstein, R. (2017). *The color of law: A forgotten history of how our government segregated America*. New York: Liveright.

Sarkar, M. (2011). *How American homes vary by the year they were built. Housing and household economic statistics*, Working Paper No. 2011–18. U.S. Census Bureau, Washington, DC.

Urban Institute. (2020). *State and local backgrounders. Housing and community development expenditures*. Retrieved from www.urban.org/policy-centers/cross-center-initiatives/state-and-local-finance-initiative/state-and-local-backgrounders/housing-and-community-development-expenditures

U.S. Census Bureau. (2023). *Housing vacancies and homeownership (CPS/HVS), Table 5*. Retrieved from www.census.gov/housing/hvs/data/detailed_tables.html

U.S. Department of Housing and Urban Development (HUD). (n.d.). *HUD's Public housing program*. Retrieved from www.hud.gov/topics/rental_assistance/phprog

U.S. Department of Housing and Urban Development (HUD). (2021). *Worst case housing needs 2021 report to Congress*. Washington, DC: U.S. Department of Housing and Urban Development, Office of Policy Development and Research.

von Hoffman, A. (2012). *History lessons for today's housing policy: The political processes of making low-income housing policy*. Joint Center for Housing Studies at Harvard University. Retrieved from www.jchs.harvard.edu/sites/default/files/w12–5_von_hoffman.pdf

Zuk, M., & Chapple, K. (2016). *Housing production, filtering and displacement: Untangling the relationships*. Institute of Government Studies, University of California, Berkeley, Research Brief.

Chapter 3

Deconstructing Affordability

Introduction

Some social outcomes are relatively easy to measure, such as poverty or infection rates. In these cases, one simply divides the number of people experiencing poverty or an infection by the total population.[1] We can compare poverty or infection rates across geographies and over time with relative ease. When it comes to housing affordability, we have the challenge of understanding two different variables: income and housing costs. Housing costs could rise, but affordability could improve if incomes rise faster than the cost of housing. Similarly, housing costs could fall, but affordability could worsen if incomes fall faster than the cost of housing. Therefore, to understand housing affordability in the United States – over time and in different locations – we need to observe and assess how housing costs and incomes interact.

In Chapter 1, we highlighted alternative measures of housing affordability, including shelter poverty and residual income. In this chapter, we focus on the most common measure, housing cost burden, which is defined as the ratio of housing costs to household income. When interpreting this ratio, researchers associate higher levels of housing cost burden with lower levels of housing affordability. In general, researchers have indicated that a household that spends more than 30% of its gross monthly income toward housing is considered cost burdened. Households that spend more than half of income on housing are severely cost burdened. The 30% threshold maps with federal housing policy as this is the percentage of income that recipients of housing assistance must pay as part of their program participation. In this chapter, we explore how housing cost burden varies by geography and over time. We also consider how changes in income and housing costs can change the relative affordability in a given community. This analysis helps to explain some rather surprising observations that emerge from the data: housing cost burdens among renters are relatively consistent across the United States and the highest cost metropolitan areas do not necessarily have the highest cost burdens.

As noted earlier in the book, the higher housing cost burdens found in Detroit when compared to San Francisco may strike many readers as surprising or implausible. How might we explain this? The low levels of housing affordability in Detroit is a function of relatively inexpensive housing costs, but even lower incomes. San Francisco tells the opposite story. Even though its housing costs are among the highest in the country, the robust labor market and the high incomes earned by residents in this city produce, on average, housing that is slightly more affordable (Myers & Park, 2022; Seymour et al., 2020). We certainly will not argue that San Francisco is a model housing market – it is far from it – but we highlight this comparison to underscore the importance of understanding both income and housing cost dynamics when analyzing the concept of affordable housing. The cases of Detroit and San Francisco are somewhat unique.

High housing costs do not always lead to lower housing cost burdens, nor do all lower cost cities or regions have high burdens.

To provide greater clarity on the relative impact of income and rents on housing affordability, researchers at the University of Southern California published a study that isolates the drivers of housing cost burden in different cities across the United States (Myers & Park, 2022). Overall levels of housing cost burden among renters is fairly consistent across the country; the level of cost burden among renters in almost all the major U.S. metropolitan areas falls within 8 percentage points of the national average. Myers and Park (2022) make an important contribution by categorizing cities based on their relative income and rent dynamics. Their typology includes metropolitan areas with:

1. growing incomes, but where rents have grown even faster (San Francisco-Oakland, San Jose, and Washington, D.C.);
2. stable incomes, where rents have still risen (Buffalo, Pittsburgh, and Miami); and
3. falling incomes, where rents have grown (Detroit, Atlanta, and Kansas City).

This study demonstrates that there are many paths to housing unaffordability. The dynamics of Detroit highlight a particularly problematic case – falling incomes and rising rents – a combination that explains the high levels of housing cost burden in a city otherwise known for its relatively cheap housing.

Housing Cost Burden Across Cities

In much of this chapter we rely on data from the U.S. Census Bureau to understand the changing relationship between income and housing costs. We begin by providing a current snapshot of housing cost burden across the country. Table 3.1 highlights the metropolitan statistical areas (MSAs) with the highest and lowest percentage of renters who are cost burdened in 2021 (to be clear, this means the percent of households that exceed the 30% housing cost burden threshold). Generally speaking, communities with the highest number of cost burdened renters tend to be places with high housing costs in California, Florida, New York, and Hawaii. These communities do not have household incomes that are sufficiently high to comfortably cover the high cost of housing. On the right side of the table, cities in the Rust Belt tend to have the lowest percentage of cost burdened renter households. The states of Pennsylvania, Ohio, Kentucky, and Indiana are well-represented on the list. Like Detroit, these communities have relatively low housing costs, but unlike Detroit, they have incomes that are sufficiently high to avoid the significant housing cost burdens experienced in that city. But it is important to note, that even in the *least* cost burdened cities in the nation, nearly 40% of all renter households are housing cost burdened. This statistic underscores that the lack of affordable housing is a major concern in all corners of the country – not just in expensive coastal cities.

In Table 3.2 we extend the analysis by looking at the housing cost burden of homeowners. Generally speaking, the burdens borne by households that own their homes are less than for renters. While the numbers are lower than what appears in the table of rental cost burdens, the story is familiar. Similar locations appear in the most and least cost burdened locations, highlighting the importance of geography when studying housing affordability (or the lack thereof) in the United States.

Table 3.1 Proportion of Renter Households Cost Burdened in 2021 by City/MSA

Most cost burdened MSAs by % of renters cost burdened		Least cost burdened MSAs by % of renters cost burdened	
Miami-Fort Lauderdale-Pompano Beach, FL Metro Area	59.6%	Dayton-Kettering, OH Metro Area	38.5%
Orlando-Kissimmee-Sanford, FL Metro Area	56.8%	Toledo, OH Metro Area	39.4%
Los Angeles-Long Beach-Anaheim, CA Metro Area	55.1%	Columbus, OH Metro Area	40.4%
San Diego-Chula Vista-Carlsbad, CA Metro Area	55.1%	Wichita, KS Metro Area	40.8%
Riverside-San Bernardino-Ontario, CA Metro Area	55.0%	Tulsa, OK Metro Area	40.9%
Urban Honolulu, HI Metro Area	54.8%	Kansas City, MO-KS Metro Area	41.4%
Las Vegas-Henderson-Paradise, NV Metro Area	54.6%	Knoxville, TN Metro Area	41.5%
Tampa-St. Petersburg-Clearwater, FL Metro Area	53.7%	Harrisburg-Carlisle, PA Metro Area	41.8%
Oxnard-Thousand Oaks-Ventura, CA Metro Area	53.5%	St. Louis, MO-IL Metro Area	41.9%
Bakersfield, CA Metro Area	53.3%	Des Moines-West Des Moines, IA Metro Area	42.2%

(*Source:* ACS, 2021)

Table 3.2 Proportion of Homeowner Households Cost Burdened in 2021 by City/MSA

Most cost burdened MSAs by % of renters cost burdened		Least cost burdened MSAs by % of renters cost burdened	
Miami-Fort Lauderdale-Pompano Beach, FL Metro Area	37.0%	Grand Rapids-Kentwood, MI Metro Area	17.3%
Los Angeles-Long Beach-Anaheim, CA Metro Area	35.6%	Harrisburg-Carlisle, PA Metro Area	17.8%
New York-Newark-Jersey City, NY-NJ-PA Metro Area	34.1%	Provo-Orem, UT Metro Area	17.9%
San Diego-Chula Vista-Carlsbad, CA Metro Area	33.8%	Knoxville, TN Metro Area	17.9%
Urban Honolulu, HI Metro Area	33.8%	Indianapolis-Carmel-Anderson, IN Metro Area	18.0%
Oxnard-Thousand Oaks-Ventura, CA Metro Area	33.0%	Cincinnati, OH-KY-IN Metro Area	18.3%
Riverside-San Bernardino-Ontario, CA Metro Area	32.5%	Kansas City, MO-KS Metro Area	18.6%
Bridgeport-Stamford-Norwalk, CT Metro Area	31.5%	Des Moines-West Des Moines, IA Metro Area	18.6%
San Francisco-Oakland-Berkeley, CA Metro Area	29.9%	Pittsburgh, PA Metro Area	18.6%
Providence-Warwick, RI-MA Metro Area	29.5%	Ogden-Clearfield, UT Metro Area	18.7%

(*Source:* ACS, 2021)

As discussed in Chapter 1, calculating housing cost burdens for owner households is more complicated than for renters. Housing cost burdens for renters are easy to calculate: one simply divides housing costs by household income. For homeowners, this is more difficult given how mortgage payments are constituted. A monthly mortgage payment includes both interest paid to the bank as well as principal payments on the loan. Because principal payments do not change the net worth of a household (they just reduce a household's outstanding debt), one can question whether the principal payment portion of a mortgage payment should be included in the housing cost calculation. Most approaches do include this in the measure of housing costs because it is a monthly cash expense for homeowners with a mortgage. Further complicating this approach are the tax benefits provided to homeowners. For many households, interest paid on a mortgage is tax deductible, which effectively reduces the true cost of a mortgage. In addition, homeowners may enjoy the benefit of increased equity value in their homes, which also is omitted from analyses of cost burden for homeowners. In the analysis in Table 3.2, housing cost burden among homeowner households account for all homeowners that spend more than 30% of their household income on housing costs (e.g., mortgage principal and interest, real estate taxes, homeowner's insurance, utilities, condominium fees).

Another way to examine variation in housing cost burden across the country is to analyze the median level of housing cost burden in each community. Rather than estimating what share of residents are cost burdened (the percentages highlighted in Tables 3.1 and 3.2), Tables 3.3 and 3.4 provide the median level of burden for renter households and homeowners. In Table 3.3, we observe a relatively narrow band between the places with the highest and lowest median housing cost burden among renters. The MSA with the highest burden is Miami and the lowest is Dayton, Ohio. The table demonstrates that even in the most affordable rental markets, median rates are close to the 30% unaffordability threshold.

Table 3.3 Median Housing Cost Burden Among Renter Households in 2021 by City/MSA

Most cost burdened MSAs by median MSA burden for renters		Least cost burdened MSAs by median MSA burden for renters	
Miami-Fort Lauderdale-Pompano Beach, FL Metro Area	37.5%	Dayton-Kettering, OH Metro Area	26.6%
New Orleans-Metairie, LA Metro Area	35.1%	Wichita, KS Metro Area	26.9%
Orlando-Kissimmee-Sanford, FL Metro Area	34.8%	Toledo, OH Metro Area	27.2%
Las Vegas-Henderson-Paradise, NV Metro Area	34.6%	Kansas City, MO-KS Metro Area	27.5%
Bakersfield, CA Metro Area	34.6%	Columbus, OH Metro Area	27.5%
Palm Bay-Melbourne-Titusville, FL Metro Area	34.6%	Harrisburg-Carlisle, PA Metro Area	27.5%
Los Angeles-Long Beach-Anaheim, CA Metro Area	34.5%	San Jose-Sunnyvale-Santa Clara, CA Metro Area	27.7%
Cape Coral-Fort Myers, FL Metro Area	34.3%	Des Moines-West Des Moines, IA Metro Area	27.8%
Baton Rouge, LA Metro Area	34.1%	St. Louis, MO-IL Metro Area	28.0%
Riverside-San Bernardino-Ontario, CA Metro Area	34.0%	Tulsa, OK Metro Area	28.0%

(*Source*: ACS, 2021)

Table 3.4 Median Housing Cost Burden among Homeowner Households in 2021 by City/MSA

Most cost burdened MSAs by median MSA burden for homeowners		Least cost burdened MSAs by median MSA burden for homeowners	
San Diego-Chula Vista-Carlsbad, CA Metro Area	22.2%	Greenville-Anderson, SC Metro Area	14.5%
Los Angeles-Long Beach-Anaheim, CA Metro Area	22.1%	Knoxville, TN Metro Area	14.8%
Oxnard-Thousand Oaks-Ventura, CA Metro Area	22.0%	Augusta-Richmond County, GA-SC Metro Area	14.9%
Miami-Fort Lauderdale-Pompano Beach, FL Metro Area	21.9%	Winston-Salem, NC Metro Area	15.0%
Riverside-San Bernardino-Ontario, CA Metro Area	21.7%	Baton Rouge, LA Metro Area	15.2%
New York-Newark-Jersey City, NY-NJ-PA Metro Area	21.5%	Jackson, MS Metro Area	15.3%
Urban Honolulu, HI Metro Area	21.4%	Akron, OH Metro Area	15.4%
Bridgeport-Stamford-Norwalk, CT Metro Area	20.8%	Pittsburgh, PA Metro Area	15.4%
Poughkeepsie-Newburgh-Middletown, NY Metro Area	20.3%	Birmingham-Hoover, AL Metro Area	15.5%
San Francisco-Oakland-Berkeley, CA Metro Area	20.2%	Toledo, OH Metro Area	15.5%

(*Source*: ACS, 2021)

Consistent with the earlier analysis, the overall level of burden among homeowners is less than for renters. The same reasons described above help to explain these differences. And again, the dynamics are similar in terms of which metropolitan areas are most and least affordable.

Housing Cost Burden Over Time

Next, we explore how housing affordability at a national level has changed over time. We begin the analysis by focusing on renter households. In 2021, the average housing cost burden of renter households in the United States was 30.6%, and the percentage of renter households that faced housing cost burden (greater than 30%) was 47.4%. A longitudinal analysis of these data show that this relationship has changed materially over the last half century. Figure 3.1 highlights that average housing cost burden has increased by about 50% and the share of cost burdened households has nearly doubled since 1970. It is important to note that for the purposes of this variable, the Census used a threshold of 35%, not 30%. Regardless of the threshold used, the key takeaway is the same: renting households today devote a far greater percentage of their monthly income to housing than households 50 years ago. This change has obvious implications for renters' ability to spend on other household necessities or save for future expenditures such as purchasing a house, college education, or retirement. When we hear stories about prior generations being able to save for a house or pay for college with far less difficulty, part of the explanation is the fundamental change in the relationship between rents and housing costs. The data clearly demonstrate that renters in 2021 are in a more precarious financial position in terms of affordability than were renters 50 years earlier.

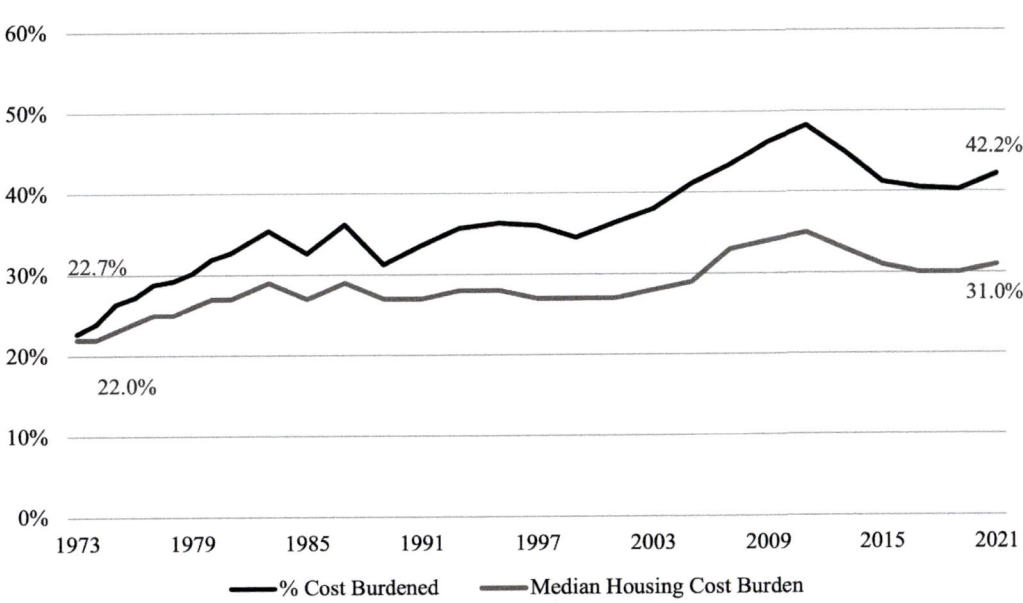

Figure 3.1 Proportion of Renters Cost Burdened (more than 35% of housing costs to income) and Median Cost Burden, 1973–2021

(Source: American Housing Survey)

We next explore the same relationship for homeowners (Figure 3.2). In 2021, the median housing cost burden among owner households was 19%, compared to 31% for renters. And the proportion of homeowners who were cost burdened was about 21%, half of the 42% of renter households that were burdened. Again, this figure is based on a 35% threshold as calculated by the U.S. Census. But in a pattern that is similar to what we observed in Figure 3.1, there has been a meaningful change over the last 50 years. The change has not been as pronounced as it has been for renters: median cost burdens for homeowners have only risen from 15% to 19%, while the percentage of cost burdened homeowner households has increased by over 40%.

While decreasing affordability among homeowners is an important issue to highlight, there is a side benefit not available to renter households. Rising home values, which may hurt affordability for potential homeowners, is a clear benefit for existing households. Over time, increases in home values become a major source of wealth creation for owners. The stark differences in household wealth between owner and renter households can, in large part, be attributed to these factors. In sum, renters have borne the brunt of the housing affordability crisis, while, in many cases, long-standing homeowners have financially benefitted from the same forces.

One of the challenges of these analyses is that averages and medians can obscure important variation in the data. As discussed earlier in this chapter, all housing cost burdens are not created equal. For example, an investment banker earning $300,000 a year who lives in Manhattan may choose to live in very expensive rental housing that costs $10,000 a month. By conventional housing affordability definitions, this person would be deemed to be housing cost burdened. But, most observers would likely agree that this person is in a very different circumstance than a household earning $30,000 that pays $1,000 a month in rent. While the rent burden fractions are identical between the two households, the implications for each household

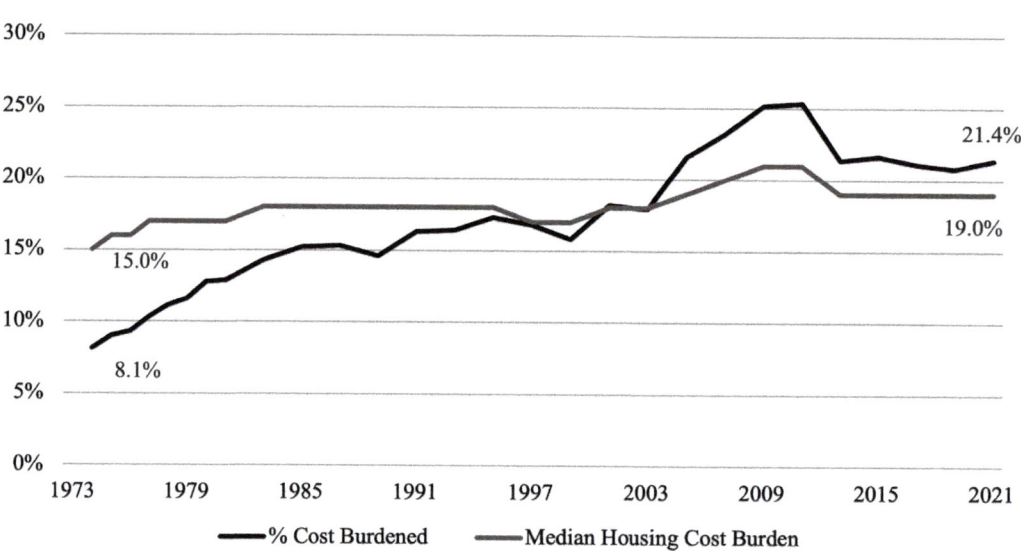

Figure 3.2 Proportion of Homeowners Cost Burdened (more than 35% of housing costs to income) and Median Cost Burden 1973–2021

(Source: American Housing Survey)

are very different. This shortcoming of the housing cost burden measure has motivated the creation of the other housing affordability measures outlined elsewhere in this book.

We conclude this section by highlighting the significant racial disparities in housing cost burden in the U.S. One can't understand and appreciate the housing system in the United States without analyzing the ways in which racist and discriminatory policies and systems shaped who could live where, who could get access to mortgage financing, and which households received the wealth benefits of increasing home equity generation after generation. Chapter 5 provides a more detailed discussion of racial disparities in housing. Figures 3.3 and 3.4 disaggregate the trends in housing cost burden by race.[2]

What emerges from these figures is the troubling and persistent racial gaps in the level of housing cost burden among U.S. households. The Black-White gap in cost burdens is between 8 to 10 percentage points, and these gaps have persisted over the last half century (Hess et al., 2022). Those who argue that racism, segregation, and discrimination are no longer issues in the modern U.S. must confront these data. Despite the passage of fair housing and civil rights legislation in the late 1960s, Black households continue to face high housing cost burdens a half century later. Obviously, the story of racial disparities in housing outcomes cannot be explained solely by the housing market. Centuries of systemic racism and discrimination have produced racial disadvantage in virtually every system in American society – education, labor market, healthcare, and criminal justice. The compounding effect of these layered systems of exclusion and oppression help to explain the significant gaps that we observe.

The analysis presented in the first portion of this chapter highlights the critical points about housing affordability in the United States. First, there has been a fundamental change in the relationship between household income and housing costs over the last half century. Households (both renters and owners) now devote a far greater percentage of their household income toward housing costs than they did 50 years ago. Not surprisingly, this change has had profound

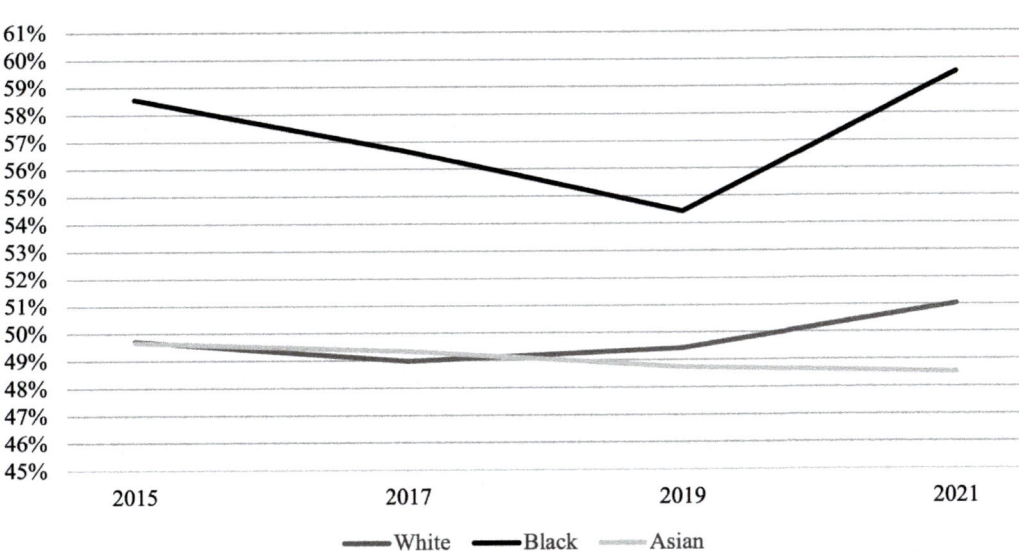

Figure 3.3 Proportion of Renters Cost Burdened by Race (more than 30% of housing costs to income), 2015–2021

(Source: American Housing Survey)

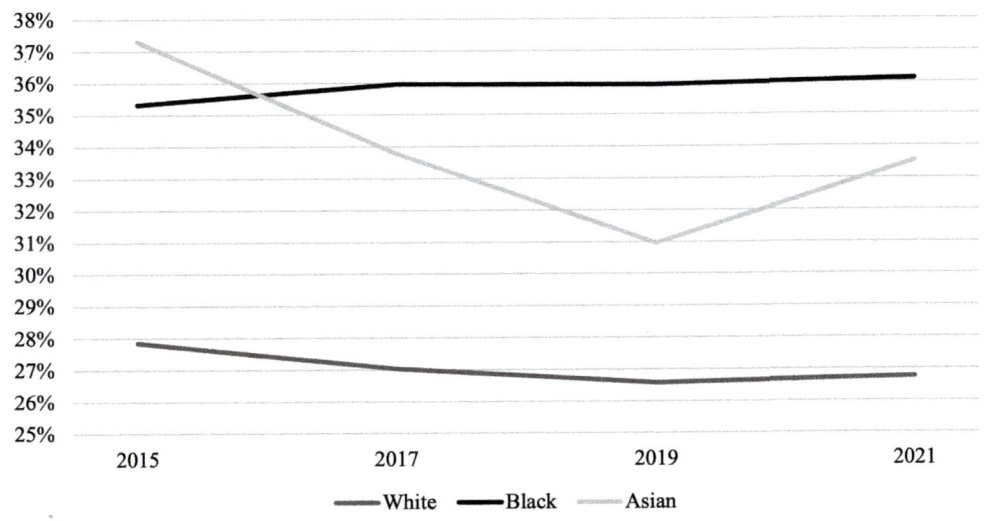

Figure 3.4 Proportion of Homeowners Cost Burdened by Race (more than 30% of housing costs to income), 2015–2021

(Source: American Housing Survey)

impacts on household budgets and has materially altered spending habits and priorities. Households today have far fewer resources to allocate to other necessary expenditures such as healthcare, transportation, education, and food (Board of Governors of the Federal Reserve System, 2023). Second, while there are varying levels of housing affordability throughout the nation, every community has a housing affordability problem. Even in the most affordable locations in the country, a significant percentage of households reside in housing that is *unaffordable* based on conventional definitions. Furthermore, households of color are disproportionately affected by the highest housing cost burdens.

Deconstructing Affordability

To shed light on the drivers of the changing dynamics of housing affordability, we use the following sections to deconstruct the housing affordability fraction. Because affordability depends on the relationship between housing costs and income, we highlight the changes in these two variables independently. These statistics highlight that the changing landscape of housing affordability in the U.S. can be explained by rapidly increasing housing costs and stagnating wages, especially among the middle and lower income households. The analysis highlights large structural trends in society that have far-reaching implications. Income inequality plays an important role in this story. Existing research highlights the dramatic rise in income inequality over the last half century (Moller et al., 2009). While overall wealth has grown rapidly over the last 50 years, gains for middle and lower income households have been paltry. Among renters, real wages (inflation adjusted) in the U.S. fell between 1990 and 2009 (Collinson, 2011). The inequitable distribution of income and wealth throughout the United States plays an important role in the growing housing affordability crisis.

The story of inequality is evident when assessing household income by tenure. Figure 3.5 highlights inflation-adjusted household income for both renters and homeowners. This figure

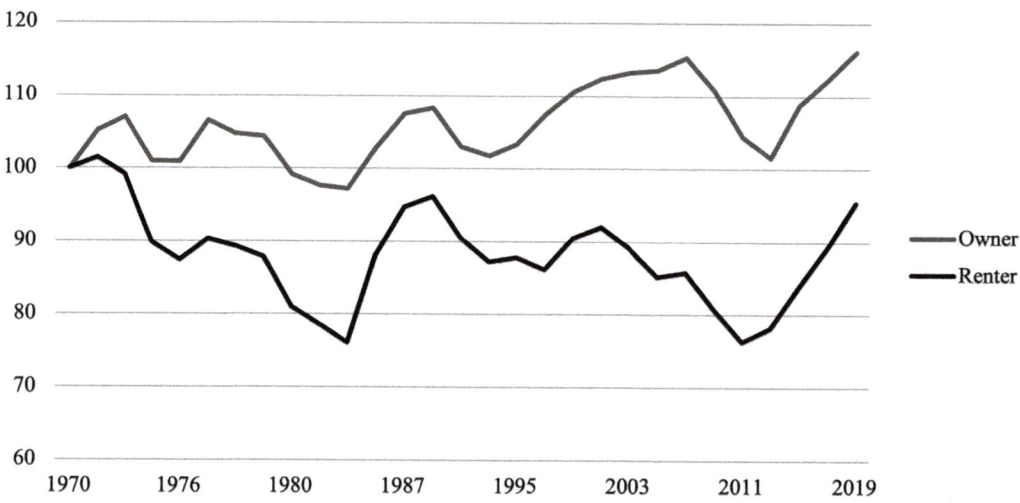

Figure 3.5 Inflation-adjusted Median Household Income by Tenure (renters vs homeowners), 1970–2019 Index 1970 = 100

(Source: American Housing Survey) Note: Data not available for all years.

shows how the incomes of homeowners have outpaced the rate of inflation in most years, while income for renters has not kept pace. This helps to explain the high levels of housing cost burden among renter households.

Compounding the challenges of modest income growth for low income, renter households, the costs of housing, both for homeowners and renters, have risen dramatically. A number of factors have driven this outcome, including population growth, low interest rates, and an inadequate supply of housing (Fernald, 2021). In many cities in the U.S., particularly on the east and west coasts, the production of housing has lagged the growth in population. Not surprisingly, the mismatch between the demand for housing and its supply has pushed the cost of housing to extremely high levels. At a national level, the inflation rate for housing costs since 1975 has exceeded many other household necessities including food, clothing, and transportation (FRED Economic Data, 1975–2021).

In Figure 3.6, we highlight these dynamics by showing the relative growth of household income and housing costs for renters over time. This graphic comes from a paper using data from the Panel Study of Income Dynamimcs (Colburn et al., 2024). Rather than showing how the levels change from year to year, Figure 3.6 shows how incomes and rents have changed across decades. This graphic highlights how the increasing cost of rental housing (after adjusting for inflation) has outpaced inflation-adjusted household incomes for renter households.[3]

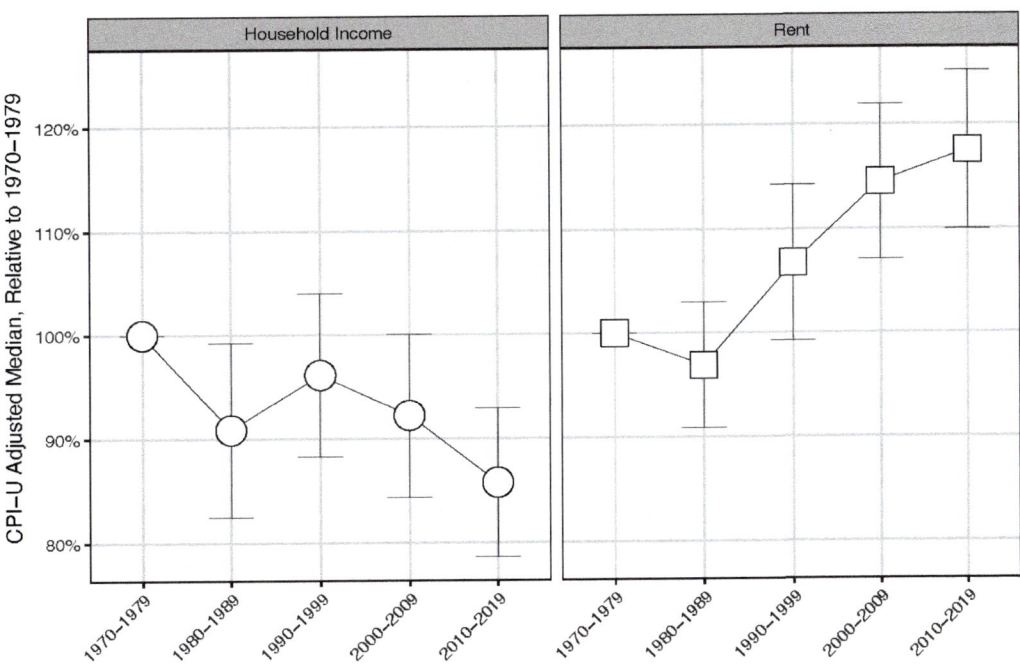

Figure 3.6 Inflation-adjusted Median Household Income and Rent for Renter Households in the U.S., 1970–2019

(Source: Panel Study of Income Dynamics; Colburn et al., 2024)

Conclusion

In this chapter, we examined affordability along multiple dimensions. First, we analyzed how housing cost burden varies across the country for both renters and homeowners. A key takeaway from this analysis is that levels of housing cost burden do not vary as much as one might expect given the dramatically different housing market conditions that exist throughout the United States. Some places have high housing cost burdens due to high housing costs, while others may have experienced cost burdens due to disproportionately low levels of household income. Since cost burden is driven by the interaction between housing costs and incomes, we must focus on both variables.

Next, we examined how housing cost burden has changed over time. What we observed is a significant increase in housing cost burden among renter households over the last 50 years. For homeowners, burdens have also increased, but not to the same extent. The implication of this finding is the conspicuous financial stress facing many U.S. households, especially those that rent. High housing costs consume a far higher percent of household income than it did two generations ago. These burdens constrain household consumption and have a range of negative consequences for people who reside in these households.

Finally, we seek to deconstruct housing affordability. We see that incomes, especially incomes for renters and those at the bottom of the income distribution have declined in real (inflation-adjusted) terms (Collinson, 2011). Over this time of stagnating incomes, rents have continued a steady climb in inflation-adjusted terms. The combined effect of these dynamics is the significant housing affordability crisis that we observe in the United States.

Notes

1 For those familiar with the study of poverty and infectious diseases, this statement is an obvious overgeneralization. There has been a lot of effort to come up with better measures of poverty and disease rates. The point of this narrative is that these outcomes rely on a single variable – how many people are in this state.
2 We use a more narrow time range for these analyses, because the data presented in Figures 3.1 and 3.2 are not disaggregated by race.
3 Figure 3.6 shows inflation-adjusted household income and rent over five decades. The bars represent 95% confidence intervals which depict the potential statistical errors in these estimates.

References

Board of Governors of the Federal Reserve System. (2023). *Economic well-being of U.S. households in 2022*. Retrieved from www.federalreserve.gov/publications/files/2022-report-economic-well-being-us-households-202305.pdf

Colburn, G., Hess, C., Allen, R., & Crowder, K. (2024). The dynamics of housing cost burden among renters in the United States. *Journal of Urban Affairs*.

Collinson, R. (2011). Rental housing affordability dynamics, 1990–2009. *Cityscape*, *13*(2), 71–104.

Fernald, M. (2021). *The state of the nation's housing 2021*. Joint Center for Housing Studies of Harvard University. Retrieved from www.jchs.harvard.edu/sites/default/files/reports/files/Harvard_JCHS_State_Nations_Housing_2021.pdf

Hess, C., Colburn, G., Crowder, K., & Allen, R. (2022). Racial disparity in exposure to housing cost burden in the United States: 1980–2017. *Housing Studies*, *37*(10), 1821–1841.

Moller, S., Alderson, A. S., & Nielsen, F. (2009). Changing patterns of income inequality in U.S. counties, 1970–2000. *American Journal of Sociology*, *114*(4), 1037–1101.

Myers, D., & Park, J. (2019). A constant quartile mismatch indicator of changing rental affordability in U.S. metropolitan areas, 2000 to 2016. *Cityscape*, *21*(1), 163–200.

Seymour, E., Endsley, K. A., & Franklin, R. S. (2020). Differential drivers of rent burden in growing and shrinking cities. *Applied Geography*, *125*, 102302.

Chapter 4

Historical Perspectives on Affordable Housing

Introduction

To understand affordable housing in the United States, one must first address the system of housing provision. The United Nations has explicitly declared a fundamental right to housing, but this goal remains aspirational in many countries around the world, including the U.S. The approach to housing provision varies dramatically around the world. For example, there is a constitutional right to housing in the Netherlands, but few other nations offer similar protections. We also see very different levels of government involvement in the construction and provision of housing. In many countries, social (or public) rental housing is over 20% of the total housing stock. These include the Netherlands (34%), Austria (24%), and Denmark (21%) (OECD Affordable Housing Database). At the other end of the spectrum are countries like the United States, Canada, and Japan where publicly-owned – or subsidized – housing constitutes less than 5% of the total housing stock (OECD Affordable Housing Database). Variation in the role of governments in housing provision has profound implications for affordability. Relying on the private market to ensure affordability may work in some contexts, but not others. Providing and sustaining affordable housing in a market-based housing system is a fundamental challenge in the U.S. and shapes much of the narrative in this book.

Larry Vale, in his book, *From the Puritans to the Projects*, highlights the legacy of weak support for subsidized housing in the United States. Public housing – housing constructed, owned, and operated by the government – is a contested topic in the United States. There has been a fundamental ideological tension between the role of the private market and the government in the provision of housing. Discussing the role of public housing in the United States, Vale writes, "At the core, the controversy about public housing is a debate about the form and purpose of state-subsidized neighborhoods in a society that places ideological value on individual homeownership and the unfettered operation of private markets" (Vale, 2007, p. 6). Public provision of housing in the U.S., writes Vale, is a "poor ideological fit" and, instead, the government has used its resources and power to enhance individual liberties and freedom that are so highly valued. Before exploring the development – or lack thereof – of the affordable housing system in the United States, it is helpful to describe how other capitalist democracies around the world have structured their housing systems. The differences, as highlighted below, are conspicuous.

International Comparisons

When analyzing national housing systems, one can focus on the relative emphasis and responsibility placed on the private market as a delivery mechanism for housing services. Scholars from multiple fields have used the notion of commodification – first developed by Karl Marx – to understand both welfare and housing systems (Forrest & Williams, 1984). Commodification, argued Marx, occurs when exchange value trumps the use value of a good or service. In the case of housing, commodification elevates the value of a home as an investment over its use value, as shelter. Commodification is at the root of a capitalist economic system. Scholars who study comparative welfare systems have used the notion of commodification to categorize nations based on the degree to which their residents are dependent on the market. Gosta Esping-Anderson, a Danish sociologist, created a typology of welfare systems based, in part, on the degree of de-commodification in a welfare system. Those nations with more generous social safety nets have a higher degree of de-commodification – meaning their citizens are *less* reliant on the private market for goods and services. Countries with high degrees of de-commodification include social democratic nations such as Sweden and the Netherlands. Liberal welfare regimes – such as the United States and the United Kingdom – have limited welfare systems and are highly commodified (Esping-Andersen, 1990). Based on the work of Esping-Andersen, housing scholars have attempted to group countries based on their housing systems. Housing systems do not match perfectly with welfare regimes (Stephens & Fitzpatrick, 2007), but housing systems can either interrupt or reinforce outcomes produced by conventional welfare policies and programs.

This brief international comparison provides important context for the understanding of affordable housing provision in the United States. Generally speaking, the government plays a relatively limited role in the provision of housing for low-income households. In this chapter, we trace the historical trajectory of housing and housing policy in the United States. It is important to understand the political, social, and economic forces that produced the current system that we observe. Through different phases of our nation's history, all levels of government have been involved in the housing system with a wide range of purposes. The government's role certainly includes a focus on ensuring housing affordability, but there have been other motivations as well.

Context for Housing Policy in the U.S.

Housing policy in the United States is not, nor has it ever been, homogeneous. These policies have served a variety of different types of households and addressed a range of different motivations. U.S. housing policy has not been solely or even primarily created to support low-income households; the ongoing unmet housing needs of low-income households speaks to the inadequacy of this policy response. There are two primary reasons for the limitations of low-income housing policy. First, even policies designed to support low-income households frequently had other motivations that, at times, undercut this support. Second, many of the housing policies enacted by the federal government were directed toward middle- and upper-income households. Low-income households were never intended to be beneficiaries of this second set of policies.

Beginning in the 1930s, the federal government took an active role in providing housing support for low-income households. This range of policies includes the public housing program, rental assistance in the form of housing vouchers, and housing production subsidies delivered as tax credits. Without question, these programs have provided necessary support to create or promote affordable housing for households in need. A closer analysis of the history highlights that these early housing policies were not exclusively focused on the lowest-income households. For example, in the early stages, public housing was built to house middle income households and the working poor; only later did public housing become housing for the lowest income households in the nation (Allen & Van Riper, 2020).

There were a range of different motivations that inspired these policies. Building public housing helped to stimulate the economy and boost employment in a time of economic hardship; for example, the public housing program was created in the 1930s to stimulate the private building industry during the Great Depression (Marcuse, 1995). The government has also used housing policy to achieve other social outcomes, including poverty deconcentration (dispersing low-income households in neighborhoods with less poverty and promoting mixed-income communities), addressing disinvestment in low-income neighborhoods, and ending cycles of poverty (McClure, 2021).

Housing policies designed to support low-income households represent only a small fraction of the total housing policies in the United States. In fact, some of the most significant housing policies in the nation's history were implemented to promote homeownership, *not* to support low-income households that are more often renters. Examples include the home mortgage interest tax deduction and government support of the mortgage finance system. To make sense of the policy landscape, we split policies into categories, which we admit, is likely an oversimplification given the complexity involved in these policies and programs: policies that support low-income households and those that promote homeownership for middle- and upper-income households. It is important to also highlight the racial implications in this legacy of policymaking. Policies to support homeownership have largely benefitted white, middle- and higher-income households. These supports have contributed to the substantial racial disparities in homeownership and wealth that we observe. On the other hand, the more limited policies designed to support low-income households have been the primary source of government support for households of color, that are disproportionately represented in the low-income population in the United States. But even these policies had potentially negative outcomes for households of color, including the concentration of affordable housing in high poverty neighborhoods. One cannot divorce federal housing policymaking from the history of racism, discrimination, and segregation in the United States.

This historical perspective informs our current understanding of the provision of affordable housing in the U.S. The housing system in the U.S., marked by heavy reliance on the private market and limited public involvement, helps to explain why the United States has fallen short of its stated goal of a decent home and suitable living environment for every American family, as initially articulated in the Housing Act of 1949. We can also position the U.S. in broader conceptual conversations about housing, including the ethical basis for a right to housing or the financialization of housing that is dominant in the United States. In sum, the United States has created a two-tiered policy system. The first tier includes policies designed to support middle- and higher-income households through homeownership, while the second tier provides residual or limited support for low-income households. One need look no further than analyses of federal spending on housing policy to see how federal resources disproportionately benefit middle and higher income households (through the mortgage finance and income tax systems) and provide limited resources (relative to total need) to low-income households. This contradiction is central to our understanding of housing policy – and particularly affordable housing policy – in the United States. Given the inadequate supply of affordable housing for the lowest income households – and the limited support from the federal government – a wide range of actors, including state and local governments and the non-profit sector, have become integral players in the attempt to provide affordable housing to those who need it.

Current Affordable Housing Landscape in the U.S.

Before delving into the history of housing policy in the U.S., it is important to understand the current landscape of housing policy and supports. The U.S. housing system can be split in two: homeownership and rental housing. The roots of this bifurcation are long-standing.

Homeownership became synonymous with the *American Dream* nearly a century ago (McCabe, 2016). Political leaders from all ideological perspectives emphasized the importance of homeownership, but such opportunities were disproportionately provided to white households. The concept of the American Dream of homeownership is highly racialized in the United States.

Homeownership has long been consistent with the American ideals of self-reliance, ownership, and independence (Dawkins, 2021). There was a prevailing belief that homeownership was good for the nation. Homeowners would have a vested interest in their communities, would form families, and pursue the American dream. Of course, renters engage in the same behaviors, but that is inconsistent with the prevailing ethos about homeownership (McCabe, 2016). In order to achieve this desired social outcome, the federal government took steps to encourage this behavior, such as creating a mortgage finance system that would provide prospective homebuyers with the financing needed to make the largest investment of their lives. The federal government also revised the tax code to encourage homeownership through the home mortgage interest tax deduction which allowed homeowners to reduce their annual taxable income. The cost of this program can be measured in lost tax revenue, rather than direct expenditure of government resources. Prior to the tax reform bill of 2017, the home mortgage interest tax deduction cost the federal government nearly $70 billion per year; an investment that dwarfed many other federal housing programs (Fischer & Huang, 2013). Post-2017, the cost of the program fell by about half (Gale, 2019) due to the more limited numbers of households that could take this benefit under the new tax rules. This long-standing emphasis on homeownership by the federal government achieved its desired outcome. Rates of homeownership in the U.S. grew from 43% at the end of the Great Depression to nearly 70% by 2004. The United States has become a nation of homeowners.

Homeownership is the primary means of wealth generation in the U.S. According to the U.S. Census, roughly 28% of household wealth in the U.S. comes from home equity (trailing only retirement accounts as the most significant source of household wealth). The wealth gap between homeowners and renters is profound. Households that own their homes have a median household wealth more than 58 times greater than that of renter households. Excluding that home equity, the gap is still 24 times (Hays & Sullivan, 2022).

Rewarding homeownership has had significant implications for society. The rewards of homeownership, by definition, only accrue to those who are able to acquire a home. Generations of race-based discrimination in housing prevented households of color from being able to acquire homes at the same rate of white households. Multiple factors contributed to this disparity – redlining, land use, and discrimination in mortgage availability (see Richard Rothstein's *The Color of Law* for a full description of this history). Research demonstrates that the large racial wealth gap can be attributed to limited access to homeownership by households of color over history. In addition, among households that have attained homeownership, the wealth benefits associated with it are lower for Black and Hispanic households compared to white households (Krivo & Kaufman, 2004). These disparities compound generation after generation.

By using the housing system to create wealth for American households, the federal government has created an internal conflict for itself. The government has used multiple tools to promote homeownership and to support the ongoing appreciation of home prices – thus expanding the wealth of homeowners. At the same time, the government has voiced a desire to ensure that housing is affordable. The contradiction in these policy goals is obvious. Despite this paradox, the federal government has continued to try to serve two masters unsuccessfully. Critics have called out the obviously hypocritical position of the government – *you can't have it both ways*. You can either promote higher housing prices, and the associated increases in household wealth, or you can fight for increased affordability. Most objective analyses would suggest that the former has trumped the latter. Homeownership has received disproportionate time, attention, and resources, and those who have been fortunate enough to own a home have been the clear beneficiaries. Those who have not had the opportunity to own a home – or chose not to – have had a very different experience and trajectory.

Historical Trajectory of Affordable Housing in the U.S.

In this section, we use two concepts to trace the development of housing policy in the United States. First, we highlight the actors and institutions that have been involved in the housing system and, second, we highlight the population of households that have been served by these policies and programs. Throughout the history of the United States, there has been an important role for local governments and non-profits in the provision of housing, particularly for low-income households. The significant functions of these players can be attributed to a couple of factors. First, housing is local and therefore local organizations and governments may be in a better position to respond to the specific needs and circumstances of a particular community. Second, local entities have responded out of necessity. Because of the relatively limited role of the federal government in providing housing support for low-income households, local governments and charitable organizations have stepped in to address the gap. While noble, this has proven difficult given the limited financial capacity of local jurisdictions when compared to the significant resources of the federal government. W. Dennis Keating described this challenge:

> Despite their best efforts, cities (and their nonprofit partners) have proven incapable of meeting the needs of all who need adequate housing. This has simply been beyond their capacity. The decline in federal housing support has limited the ability of these local providers to meet increasing housing needs.
>
> (Keating, 2021)

The concept of a two-tiered housing system appears throughout this narrative, both in this and upcoming chapters. The first tier traces its roots to the beginning of the 20th century and was designed to support the private housing market and promote homeownership. These policies were designed to support business groups and institutions through subsidization from the federal government. The second tier began with the Housing Act of 1937. This act was focused on developing public housing that initially served middle income households, but subsequently became housing for the highest need households in the nation. The program was influenced by the Modern Housing proposal popularized by famous "houser", Catherine Bauer, but the act fell short of what the housers had hoped for. Bauer advocated for multifamily housing and neighborhoods accessible to people of all incomes and constructed to mask individual wealth, similar to the current mixed-income housing and neighborhood approach. Bauer knew housing solely built for the poor would not gain support and thus lack the necessary resources to create and maintain a decent living environment (Bauer, 2020). Bauer's insights proved to be accurate, as efforts to construct public housing were met with opposition from business and real estate interests because of fear that government involvement in housing would compete with the private market. Throughout history, the government has sought to fulfill the mission of the second tier with incomplete efforts which were complicated by the ongoing desire to avoid disrupting the private market for housing. This two-tiered system has also been racialized. The beneficiaries of the first tier are largely white and affluent, while those helped by the second tier of policymaking have been disproportionately poor and people of color. We discuss racism, discrimination, and segregation in the housing market in greater detail in Chapter 5.

Present Day

While not a linear approach, we begin our journey with a presentation of the current state of the housing system in the U.S. From there, readers can then follow the trajectory of U.S. housing policy with this end in mind. Our current approach to providing varying levels of support to households in different situations is not an accident but rather the result of a concerted effort to protect private interests and the private housing market over the last 250 years of our nation's history.

The current system is a patchwork of federal, state, and local policies combined with a range of programs and services provided by non-profit organizations. As noted previously, at the federal level, resources disproportionately benefit homeowners:

> Of the $152.1 billion in federal tax expenditures in fiscal year 2012, 86%, or $130.2 billion, went to homeowners. By far, the largest tax break is the deductibility of mortgage interest payments from taxable income. These deductions accounted for 54% of all housing-related tax expenditures in 2012 and 63% of all homeowner tax expenditures.
>
> (Schwartz, 2014, p. 117)

Following the tax reform of 2017, the dollars allocated toward homeowners has fallen, but the glaring inconsistency of federal support for homeowners and renters persists. These stark statistics demonstrate that the primary purpose and intent of federal housing policy has been, and continues to be, to support and encourage homeownership, and to create a system that increases the values of homes. Supporting low-income households has always been a secondary goal, based on the level of expenditures allocated to each tier. This speaks to the fundamental disconnect in federal housing policy. If the majority of a government's policies exist to support and increase home values, this approach is in direct conflict with the goal of ensuring or providing affordable housing. The government has been unequivocally successful with the former goal, and substantially failed in the latter. A robust change to the housing system in the U.S. will require a national reckoning in which we acknowledge this disconnect and chart a different path in which the resources, energy, and focus of the U.S. government is shifted to the second tier. To date, the political will and courage needed to make such a change has not materialized.

In the absence of a robust federal response, states, localities and the non-profit sector have emerged to play a significant role in the U.S. housing system. And much, although not all, of these efforts have been to support the second tier of housing supports. Many non-profit organizations have been created to provide housing, shelter, and services to people in need. Many of these programs were created by churches and religious organizations. Over time, the sector has become more secular, but their efforts continue to fill the gaps created by inadequate federal policymaking. For example, community development corporations (CDCs), which are non-profit organizations that support community development efforts, emphasize affordable housing development and there are now nearly 5,000 CDCs in the nation. The non-profit sector is particularly notable in the homeless response system.

States and local jurisdictions also have stepped into the breach. A range of programs including rental assistance, production subsidies, down payment assistance, foreclosure prevention, and financial literacy and homeownership education have been created and funded to support low-income households. While some of this local and state funding flows from the federal government, many of these programs are funded with state and local resources. Direct general expenditures fund housing and community development activities but since there are many competing needs such as public infrastructure and schools, these activities usually only comprise less than 2% of all state and local direct general expenditures (Urban Institute, 2020). Housing advocates contend that dedicated sources of funding, such as housing trust funds, are essential to support affordable housing initiatives. Almost all states and many large cities have established housing trust funds with dedicated sources from real estate transfer taxes, document recording fees, revenue from state housing finance agencies, or other sources such as state income tax or bond proceeds. Although difficult to quantify since housing trust funds are a patchwork of programs, it is estimated that state housing trust funds alone generated over $1.6 billion for housing initiatives in 2020 (Housing Trust Fund Project, 2021). Finally, states and localities have also partnered with the federal government to promote the first tier of housing policymaking. Notably, the disproportionate reliance on single-family zoning has enhanced the value of owner-occupied housing in many communities throughout the United States. In addition, with property tax and homestead tax exemptions, local jurisdictions further support homeownership, which is discussed in detail in Chapter 10.

We now begin our historical journey of the U.S. housing system. As you read, consider the current two-tiered housing system and the disproportionate resources and attention devoted to each tier. This history highlights that the current shortcomings in the U.S. policy response cannot be attributed to one era or a single politician. Rather, the creation of the present system can be traced back to the founding of this nation, was expanded and sustained over time, and reinforced in current housing policies. The longer the current system continues to exist, the greater its inertia that prevents meaningful change.

Colonial Era to 19th Century

Since its founding, the United States has struggled with the challenge of meeting the needs of people who have been unable to maintain stable housing themselves. Tracing this legacy to the current day, Larry Vale (2007) writes:

> Since the time of the seventeenth-century Puritan villages, American communities have struggled to define the nature and extent of their ethical and jurisdictional responsibilities to public neighbors – those judged socioeconomically unable or unwilling to meet their community's standards of industry or behavior. The term *public neighbor* simultaneously encodes both social obligation and spatial proximity. Over the centuries, these matters of obligation and proximity have found expression in many different institutional and architectural forms – from almshouses to model tenements to settlement houses to public housing projects – and the jurisdictional and ethical issues have inevitably become more complex and more contested.
>
> (p. 19)

From the Colonial Era through the mid-1800s, the United States was marked by a strong emphasis on local control. Communities largely operated autonomously with limited support from larger or higher levels of government. Because the economy was disproportionately focused on agriculture, urban population density was relatively low, and the future challenges associated with urban life during the industrial revolution had yet to materialize. One could think about the housing system in this era as almost entirely local and privately delivered.

As many scholars have highlighted, the approach to need during this time was situationally-dependent. Support was only provided to those considered "local," strangers from other villages or communities would not receive support or assistance. This parochial approach to welfare provision echoes to the current day debates about public assistance in the United States. A second key theme is that of deservedness. Public obligation extended only to those deemed deserving based on their life circumstance – widows, orphans, and those with disabilities. Able-bodied poor people (mostly men) who demonstrated need were considered *undeserving* and therefore received limited support. This approach was consistent with the Elizabethan poor laws that the Puritans brought with them to America.

In this era, there was a heavy emphasis on private responses to need. The family and the church played an important role in providing care and lodging for individuals in need. These private sources were the first line of defense. Outdoor relief was intended to keep people in their homes so that they didn't become the responsibility of the community. In cases where these private sources of support were inadequate or ineffective, the public response was indoor relief which included accommodations such as poorhouses, almshouses and workhouses (Jensen, 2015; Katz, 1996; Vale, 2007). This housing was provided by the local government and was the last resort for people in need. Workhouses required labor from its residents and were reserved for the undeserving poor, while almshouses provided lodging without a work requirement – for those who had a "legitimate" reason for not working.

While much local attention was paid to the poor during this era, the fledgling federal government was hard at work promoting the ideals of individualism, freedom, and upward social and economic mobility (Vale, 2007). The Land Ordinance of 1785 established the system of selling public land to individual households that could afford it. Although there were not any explicit restrictions based on race or gender, people of color and women were less likely to benefit because of barriers to land ownership for them at that time. This was the beginning of the federal government's long-standing effort to support and encourage homeownership. Nearly a century later, the Homestead Act of 1862 provided land to households with the only requirement that they live on the land. According to the National Association of Realtors, this act, "was the beginning of the concept of the American Dream of owning a home for your family" (Sanfilippo, 2018). In this era, the federal government focused on land and homeownership, the importance of families, and individualism, while local governments were left to address issues related to poverty and housing precarity.

As the United States began to urbanize and the industrial revolution brought change to society, there was a need to house the increasing number of urban dwellers. In addition, waves of immigrants poured into U.S. cities – notably New York City – and needed places to live. These households sought housing in a system that was, at that time, largely unregulated. For the latter half of the 19th century, tenements were the primary form of housing for many working-class urban households. Tenements were densely constructed with little focus on green space, ventilation, or sanitation. Jacob Riis' famous book, *How the Other Half Lives*, highlighted the plight of those living in tenements, who were largely hidden from the view of middle- and upper-class urban residents (Riis, 1890). The poor conditions found in tenements prompted the first housing reform law in the nation's history. In 1867, New York State passed the Tenement House Law which established minimum standards for housing quality, room size, ventilation, and sanitation. Tenements remained a primary source of housing until the reforms of the New Deal changed the trajectory of housing in America. This period marks how state governments began to take an active role in providing oversight of the private housing market.

Progressive Era to Great Depression

The onset of the Progressive Era in the 1880s ushered in a new approach to housing in the U.S. During this time, there was increased action by both local and federal governments in the domain of housing. In response to growing urban challenges, public scrutiny centered on substandard housing and tenements. We observe increased regulatory action by states, notably the State of New York. A series of laws designed to regulate the poor conditions of tenements culminated with the New York State Tenement House Act of 1901, which banned the construction of tenement-style housing. The law followed prior legislation that sought to improve the safety of tenements by requiring fire escapes, indoor plumbing, and access to fresh air. These efforts highlighted how the government could play an important role in the safety and quality of the residential built environment. The advances were driven by advocacy from Progressives who sought a better life for the growing population of urban dwellers.

The decades between the Progressive Era and the Great Depression ushered in a new and prominent role of the federal government in the housing system. The federal government was a new entrant in the system and its impacts were significant. Local jurisdictions no longer assumed primary responsibility for housing and the federal government began playing a more meaningful role. There were three notable events or drivers that marked the growing involvement of the federal government. First, was the severe housing shortage that existed after the end of World War I. Housing production had slowed as resources and attention were directed to the war effort. In addition, there was growing demand for the scarce housing as rural transplants and immigrants flocked to U.S. cities (Claire, 2020). This shortage prompted the federal government – which had previously remained on the sideline – to become an active participant in the nation's housing system. And, rather than using the apparatus of government to address the shortage of housing, the federal government sought partnership with private industry to address this need.[1] Two significant programs were developed, the Better Homes in America program and the Own Your Own Home program. These initiatives were designed to encourage suburban homeownership for (largely white) urban dwellers. The programs also supported and encouraged private development of housing, without having to resort to publicly-funded construction of housing. Vincent J. Cannato described this period:

> It was around this time, and especially after the First World War, that the belief in the social value of homeownership first found expression in public policy. Federal support began as an extension of anti-communist efforts in the wake of the Bolshevik Revolution in Russia; as one organization of realtors put it at the time, "socialism and communism do not take root in the ranks of those who have their feet firmly embedded in the soil of America through homeownership." A public-relations campaign dubbed "Own Your Own Home" – originally launched by the National Association of Real Estate Boards in the aftermath of World War I – was taken over by the U.S. Department of Labor in 1917, and became the first federal program explicitly aimed at encouraging homeownership.
>
> (Cannato, 2023)

As seen in the Cannato quote, the second driver of the federal government's role in housing was ideological in nature. The government sought to not only address the housing shortage but also achieve ideological ends in doing so. Herbert Hoover played a prominent role in these efforts, both as Secretary of Commerce under Presidents Harding and Coolidge and later after being elected President himself. Hoover was the architect of the Better Homes in America program which overtly encouraged white, middle-class homeowners to establish the right kind of homes in the right kind of communities (i.e., in the suburbs). The hope and intent was to instill a culture of ownership and civic responsibility among this class of households. Hoover's goal was to make ownership of single-family homes in zoned, planned communities the focus of American housing policy and he adopted zoning regulations and standardized building codes toward that goal. Through all these endeavors, Hoover promoted the role of private capital and the real estate industry in crafting a federal housing policy.

In his 1922 book, *American Individualism,* Hoover connected homeownership with independence and initiative, explaining that Americans needed to own homes as part of their national identity (Hoover, 1922). He worried that insufficient housing would create conflict between two classes: tenants and landlords. Hoover thought home construction was a key to stimulating the economic recovery and would also help eliminate barriers to homeownership. The Better Homes in America program was intended to solve American's housing problems without direct government funding of housing. The program celebrated homeownership, home maintenance and improvement, and home decoration as means of motivating responsible consumer behavior; it also expanded the market for consumer products. Annual local campaigns encouraged people to own, build, remodel, and improve their homes and distributed advice on creating home furnishings and decorations. Hoover believed that a healthy construction industry served as both a base for and an indicator of U.S. economic strength.

Soon after Hoover was elected President in 1929, the Great Depression set in. The economic hardship associated with the depression had profoundly negative consequences for the nation's new class of homeowners. Unemployment rose to 25% and millions of people were no longer able to make their mortgage payments. Foreclosures were prevalent and banks failed at alarming rates. The entire housing system was in shambles. For those unable to find or maintain housing during this period, large encampments of people experiencing homelessness popped up in cities throughout the nation. These communities were derisively referred to as "Hoovervilles" after the nation's President who led the response to the economic crisis.

Franklin D. Roosevelt was elected to the Presidency in 1932 in the depths of the Great Depression. There were immense needs associated with the economic downturn. Roosevelt needed to stimulate the economy, stabilize the banking system, and repair the nation's housing system. During the decade of the 1930's, Roosevelt's administration laid the foundation for the modern housing finance system and also created a role for the federal government in the provision of housing for low- and middle-income households. The legacy of these important programs is still evident today.

These events and the economic hardship and dislocation associated with the Great Depression were the catalysts for the third driver of federal involvement in the U.S. housing system. One of the most obvious consequences of the economic crisis was the inability of homeowners to make their mortgage payments. By 1933, one half of all mortgages were in default. Alex Schwartz described the federal government's response to the foreclosure crisis and the effects of these programs over the subsequent decades:

> Faced with widespread mortgage foreclosure and the collapse of the entire housing industry, the federal government responded with a series of initiatives that utterly transformed the nation's housing finance system and helped propel homeownership within reach of a majority of its households. These programs and institutions paved the way for the nation's remarkable increase in homeownership from the 1940s to the 1960s and established a new, stable system for housing finance that stood solid for more than 40 years.
>
> (Schwartz, 2014, p. 69)

One example of these efforts was the Home Owner's Loan Corporation (HOLC) which sought to provide mortgage relief to troubled homeowners. In addition, the HOLC laid the groundwork for the Housing Act of 1934 which created a comprehensive housing plan for the nation. Importantly, the act created the Federal Housing Administration (FHA), which provided insurance to banks and lenders, in turn helping to promote and encourage the construction of new housing. These are two examples of the range of policies established by the federal government to strengthen the banking system and support mortgage lending to U.S. households. As will be discussed further in Chapter 5, the practice of redlining began with a system of appraisals that privileged homes in white neighborhoods. These policies provided the financial and institutional stability needed to achieve the ideological goals of a creating a homeownership society in the United States – specifically a homeownership society for white households. The post-World War II realization of vast suburbanization and the "white picket fence" American Dream are the direct result of these legislative efforts.

In addition to shoring up the housing finance system, the federal government also took a major step to change the way in which affordable housing was provided. Prior federal efforts were exclusively focused on the homeownership market and largely served middle- and higher-income homeowners. Recognizing the huge unmet need evident during the 1930s, President Roosevelt noted in his second inaugural address, "I see one-third of a nation ill-housed, ill-clad, and ill-nourished" (Roosevelt, 1937). Housing advocates, such as Catherine Bauer of the Regional Planning Association of America, pushed aggressively for publicly-funded and constructed housing, i.e., a public housing program. In response to the demonstrable need as well as the economic conditions of the Great Depression, Congress passed and President Roosevelt signed the Wagner-Steagall Housing Act in 1937. This legislation established the United States Housing Authority which provided funding to local housing authorities to construct public housing. According to the legislation, the purpose of the bill was,

> To provide financial assistance to [state and local governments] for the elimination of unsafe and unsanitary housing conditions, for the eradication of slums, for the provision of decent, safe, and sanitary dwellings for families of low income, and for the reduction of unemployment and the stimulation of business activity, to create a United States Housing Authority, and for other purposes.

As this description highlights, economic growth and stimulus were important motivators alongside housing affordability. Echoes of this dual motivation have played out throughout the history of housing policy in the U.S. Helping poor households has always been more palatable if the nation can stimulate the economy while doing so.

World War II to 1970s

Housing policy in the U.S. continued to develop over the remainder of the 20th century. First, the end of World War II and the housing shortages associated with with the large numbers of returning soldiers, necessitated significant housing development. This development came both in the form of private development of housing plus the expansion of the construction of public housing projects throughout the country. In addition, a range of housing finance agencies were established, as was a Veteran loan program for returning soldiers. The 1949 Housing Act outlined a goal of "a decent home in a suitable living environment for every American family." This stated goal marked a shift in federal thinking about housing from an isolated problem to a necessary component of a well-functioning society. This logic is evident in the urban renewal movement of the 1950s that sought to eradicate the urban slums that had become prevalent in many

U.S. cities by constructing new public housing in their place or other commercial development that would increase the city's tax base. In the 1960s, the Department of Housing and Urban Development (HUD) was established and the Civil Rights Act and Fair Housing Act were adopted. The significant construction of public housing in the 1950s and 1960s had its roots in the New Deal legislation of the 1930s. The 1970s brought the Housing and Community Development Act of 1974 and the beginning of the end of the public housing program in the United States. The scope and trajectory of policymaking over these four decades was notable. These initiatives galvanized a host of homeownership, rental, and community development programs and had a significant impact on the residential built environment in the United States.

The 1949 Housing Act provided billions of dollars for slum clearance and the construction of public housing. Ultimately, the full potential of these efforts was not achieved. In addition, these parallel activities – slum clearance and public housing construction – perpetuated and accelerated racial segregation (Rothstein, 2017). Ambitious targets for public housing construction fell short, but urban renewal demolition was robust. Millions of units of housing were constructed, but even more units were lost through urban renewal demolition. Greg Miller (2017) described this activity:

> Urban renewal projects changed the landscape of American cities in the 1950s and 1960s. The federal government gave cities billions of dollars to tear down blighted areas and replace them with affordable housing. Or at least, that's what was supposed to happen. In many places there was a net loss of housing as city leaders decided instead to build offices or shopping malls, or to expand hospitals and universities.

The net effect of these efforts was a loss of housing, particularly for poor, Black households. Scholars from the University of Richmond estimate that over 300,000 people were displaced in just over a decade from the mid-1950s to the mid-1960s (Digital Scholarship Lab, 2023). The scale of public housing construction, which provided new, quality, housing for low-income households, proved to be inadequate relative to the demonstrated need of this era. The inadequate federal response to housing need is a theme that repeats throughout the twentieth century.

Beginning in the 1970s, public housing went through a transition. In its original form, public housing was not designed to serve households with the lowest incomes. Rather, it was intended to serve working class households who were experiencing hardship associated with the Great Depression. Over time, the profile of public housing residents changed; the program began to serve increasingly impoverished tenants. While serving those with the greatest needs made sense, it put significant financial pressure on housing projects. Because a portion of the operating revenue for the building came from rent paid by tenants, as the income of residents fell, so did rental revenue needed to maintain and operate the buildings. In response, local housing authorities began to increase the rents they charged public housing tenants. These changes prompted protests from tenants and there was a growing concern that tenants must be protected from increasing rents. The legislative response came in the form of the Housing and Urban Development Act of 1969, which included the "Brooke Amendment." The Amendment set a cap for the percentage of income that could be charged as rent in public housing. The original limit was set at 25% and was later increased to 30%. While well-intentioned, the rent cap combined with ever-increasingly poor tenants cemented the financial challenges of public housing. Since the federal government chose not to increase their contribution for operating costs and maintenance, much of the nation's public housing stock fell into disrepair over the succeeding decades as rent revenue was insufficient to operate and maintain the units at an adequate level. The declining quality of the housing stock contributed, in part, to its demise. But, as Edward Goetz writes in *New Deal Ruins*, the decline of public housing wasn't inevitable, rather it was a policy choice that has had significant consequences on the affordable housing landscape in the U.S. (Goetz, 2013). Goetz argues that much of the public housing stock, had it been maintained and operated properly, could have continued to provide adequate housing to millions of U.S. households.

Beginning in 1973, ideological differences emerged in the nation's approach to low-income housing policy. That year, George Romney, then Secretary of HUD, and President Nixon ordered a moratorium on all new construction of public and subsidized housing. Budget authority for housing programs declined throughout the duration of the decade. This period marked the beginning of the end of the public housing era in the United States. On the heels of the moratorium, the Housing and Community Development Act of 1974 was passed which established the Section 8 (housing voucher) program. This legislation was significant because it marked the transition from publicly-financed and operated housing programs, to subsidy programs that allowed tenants to access private market housing with the benefit of a federal subsidy or voucher. In essence, the new approach used the private housing market as a delivery mechanism for public benefits. There were two potential motivations for this transition. First, it was consistent with the broader political movement that incorporated neoliberal ideology into a range of public policymaking. Neoliberalism favors private market delivery of goods and services over public provision, so the shift to vouchers is entirely consistent with this ideology. Second, there was a broad understanding that in contrast to the post-World War II period when housing access and housing quality were important social concerns, the housing challenges of the 1970s were more about affordability than quality. Therefore, it was argued, a shift from housing production to providing purchasing power to low-income households was the solution. In reality, these two explanations are complementary and together they played a role in this significant policy transition.

This era also marked the beginning of more sustained action by state governments as well as non-profit providers of housing. For the first 200 years of U.S. history, state governments played a relatively limited role in the development and provision of affordable housing. This began to change in the 1960s when the first state housing finance agency (HFA) was created in New York. Within 20 years, 42 states had created their own HFA (Williams, 2021); these entities became important sources of state and local government involvement in affordable housing. Other new actors in the housing system were non-profits, which were officially invited to participate in the provision of federal housing programs in 1959 with the Section 202 low-interest loan program for elderly housing and in 1961 with the Below Market Interest Rate program (O'Regan & Quigley, 2000). CDCs, introduced in the 1950s, became significant contributors to the affordable housing stock (Bratt, 2021). CDCs entered the picture in response to civil unrest and urban renewal and became the largest contributor to the nonprofit housing stock in response to the ongoing conspicuous housing challenges of low-income communities throughout the country. Their housing production activity accelerated during the 1980s and beyond and within 4 decades CDCs produced more than 400,000 affordable housing units (O'Regan & Quigley, 2000).

1980s to Present

The current landscape of affordable housing policy in the U.S. has its roots in the transition that began in the 1970s and accelerated in the 1980s. Consistent with the neoliberal turn in policymaking worldwide, housing policy in the U.S. began to emphasize the importance of markets in all domains of life. Gone were the days where government provision of housing was the dominant approach to meeting the needs of low-income households. Throughout the 1980s, there was a conspicuous federal retrenchment from housing programs and policies. In its place came market-oriented approaches such as housing vouchers and tax credits. These programs remain, today, as the most significant ways in which the federal government supports low-income households (with vouchers) and promotes the construction of affordable housing for low-income households (through the issuance of federal tax credits).

This transition had real human and social costs. The shift away from public housing – and the subsequent demolition of millions of public housing units in this era – led to significant displacement and hardship for the former tenants of these buildings. In their place, new mixed income developments were built (with support from the federal HOPE VI program), but the number of affordable, replacement units was relatively small when compared to the number of units that were lost. From 1991 to the present, over 370,000 million units of public housing were torn down in the U.S. By 2022, the U.S. operated a public housing portfolio of just under a million units, down from its peak of nearly 1.4 million units in 1991.

With the relative demise of the public housing program, a new mechanism to support the production of affordable housing was created. As part of Ronald Reagan's 1986 Tax Reform Act, the Low Income Housing Tax Credit (LIHTC) was established. The purpose of this program was to provide tax credits to help fund the development of housing that is affordable to households with lower incomes. Rather than funding, constructing, and operating public housing, the federal government now could incentivize the construction of affordable housing by issuing tax credits to private entities. With the support of LIHTC, both for-profit and non-profit housing developers construct affordable housing. In exchange for receipt of tax credits, the developers agree to construct housing that is affordable to households with specified levels of income for at least 30 years. According to HUD, between 1987 and 2021, LIHTC has supported the development of 3.55 million affordable housing units. As of 2022, there are 2.3 million units in operation (Acosta, 2022). While considerable, this production of housing has not kept up with need and the program has not been effective at serving the lowest income households with the greatest needs (Williamson, 2011).

Another significant source of affordable housing that was developed in this period was housing constructed by nonprofit agencies. According to Bratt (2021), there are nearly 3.5 million affordable housing units developed and operated by nonprofit providers. Given that there is a limited supply of housing that is subsidized by the government, this contribution from the non-profit sector substantially increases the supply of de-commodified housing – housing that is outside of the private market. There are a variety of different kinds of non-profit housing providers, including CDCs, members of the Housing Partnership Network, community land trusts, limited equity cooperatives, and mutual housing associations. The growth of this sector during the 1990s was substantial; in 1990, there were approximately 2,000 CDCs and by 1997 that number had grown to 3,600 (Bratt, 2021). The best, most recent figure suggests that there are now 4,600 such organizations in the United States (National Alliance of Community Economic Development Associations, 2014). The growth of non-profit CDCs can be attributed to three national community organizations: NeighborWorks America, Local Initiatives Support Corporation, and Enterprise Community Partners (Bratt, 2021).

This era also includes the expansion of state and local governments' involvement in the provision of affordable housing. As described by Victor Basolo, the main activities of state-initiated housing policies and programs include: revenue generation to fund the provision of affordable housing, actual affordable housing programs offered by the state, and regulatory actions (at the state level) to promote affordable housing (Basolo, 2021). Examples of these programs include state housing finance agencies which began in the 1960s and flourished after the creation of the LIHTC program. In addition, many states started housing trust funds that generate money dedicated to fund affordable housing activities without having to go through an appropriation process. Sources for state housing trust funds include real estate transfer taxes, document recording fees, revenues from state housing finance agencies, interest from real estate escrow accounts, bond proceeds, and state income taxes (Center for Community Change, 2016). As of 2011, 47 states had a housing trust fund. Many states have also responded to the demand for tax credits provided by the LIHTC program by establishing their own tax credit programs. These efforts mimic the federal program but use state funds (tax credits) to expand the resources within the state that can be used to construct affordable housing. About one third of all states have their own tax credit program (Basolo, 2021). States have also exerted their power and authority to promote affordable housing by requiring localities to create comprehensive plans for housing. Nearly half of all states have pursued this approach (Ramsey-Musolf, 2017). These actions demonstrate that states are not powerless when it comes to the provision of affordable housing. Given the inadequacy of federal efforts, states have stepped in to, in part, address the gap.

This historical perspective informs our current understanding of the provision of affordable housing in the U.S. The housing system in the U.S., marked by heavy reliance on the private market and limited public involvement, helps to explain why – in the opinion of many scholars and observers – the United States has fallen short of its stated goal of a decent home and suitable living environment for every American family. We can also position the U.S. in broader conceptual conversations about housing, including the ethical basis for a right to housing and the financialization of housing that is dominant in the United States. In sum, the United States has created a two-tiered policy system. The first tier includes policies designed to support middle- and higher-income households through homeownership, while the second tier provides residual or limited support for low-income households. One need to look no further than analyses of federal spending on housing policy to see how federal resources disproportionately benefit middle- and higher-income households (through the mortgage finance and income tax system) to the detriment of low-income households with greater needs (Fischer & Sard, 2017). This chapter highlights the uneven allocation of housing supports. The inadequate affordable housing resources for the lowest income households, particularly from the federal government, have prompted a wide range of actors at the state and local level to participate in the complex process of providing affordable housing to those who need it.

Conclusion

As we move from the history detailed in this chapter back to the present day, we might ask: Has low-income housing policy in the U.S. been a success or failure? Alex Schwartz addresses this question on the opening page of his book, *Housing Policy in the United States*. Since the passage of the 1949 Housing Act, Schwartz writes, "the federal government has helped fund the construction and rehabilitation of more than 5 million housing units for low-income households and provided rental vouchers to nearly 2 million additional families. Yet, the nation's housing problems remain acute" (Schwartz, 2014, p. 1). Supporting millions of low-income households can certainly be held up as a major achievement, but the inadequacy of the response – in the face of a large, demonstrable need – suggests that these federal efforts have been ineffective, or more accurately, insufficient. In response to the need, states, cities, and non-profits have attempted to fill the gap with varying levels of success. In sum, affordable housing policymaking and programs in the U.S. can best be described as incomplete.

Note

1 Manisha Claire wrote an excellent summary of this history. Claire, M. (2020). The latent racism of the Better Homes in America Program. *JStor Daily*, February 26, 2020. Retrieved from https://daily.jstor.org/the-latent-racism-of-the-better-homes-in-america-program/#:~:text=By%201923%2C%20a%20Los%20Angeles,fix%20to%20its%20housing%20woes

References

Acosta, S. (2022). *Final 2023 funding bill should support, expand housing vouchers*. Washington, DC: Center on Budget and Policy Priorities. Retrieved from www.cbpp.org/blog/final-2023-funding-bill-should-support-expand-housing-vouchers.

Allen, R., & Van Riper, D. (2020). The New Deal, the deserving poor, and the first public housing residents in New York City. *Social Science History*, 44, 91–115.

Basolo, V. (2021). Affordable housing: Program financing and policies in the U.S. States. In K. B. Anacker, M. T. Nguyen, & D. P. Vardy (Eds), *The Routledge Handbook of Housing Policy and Planning* (pp. 390–405). New York: Routledge.

Bauer, C. (2020). *Modern housing*. Minneapolis, MN: University of Minnesota Press.

Bratt, R. (2021). The U.S. approach to social housing. In K. B. Anacker, M. T. Nguyen, & D. P. Vardy (Eds), *The Routledge Handbook of Housing Policy and Planning* (pp. 173–188). New York: Routledge.

Cannato, V. (2023). A home of one's own. *National Affairs*, 56(Summer). Retrieved from www.nationalaffairs.com/publications/detail/a-home-of-ones-own

Center for Community Change. (2016). *The 2016 housing trust fund survey report*. Washington, DC: Center for Community Change. Retrieved from www.housingtrustfundproject.org

Claire, M. (2020). The latest racism of the Better Homes in America Program. *JSTOR Daily*, February 26, 2020. Retrieved from https://daily.jstor.org/the-latest-racism-of-the-better-homes-in-america-program/#:~:text=By%201923%2C%20a%20Los%20Angeles,fix%20to%20its%20housing%20woes.

Dawkins, C. J. (2021). *Just Housing: The Moral Foundations of American Housing Policy*. Cambridge, MA: MIT Press.

Digital Scholarship Lab. "Renewing inequality," *American Panorama*, ed. Robert K. Nelson and Edward L. Ayers. Retrieved from https://dsl.richmond.edu/panorama/renewal/#view=0/0/1&viz=cartogram&text=about

Esping-Andersen, G. (1990). *Three Worlds of Welfare Capitalism*. Princeton University Press: Princeton, NJ.

Fischer, W., & Huang, C. (2013). *Mortgage interest deduction is ripe for reform*. Washington, DC: Center on Budget and Policy Priorities.

Fischer, W., & Sard, B. (2017). *Chart book: Federal housing spending is poorly matched to need*. Washington, DC: Center on Budget and Policy Priorities.

Forrest, R., & Williams, P. (1984). Commodification and housing: Emerging issues and contradictions. *Environment and Planning A: Economy and Space*, 16(9), 1163–1180.

Gale, W. G. (2019). *Chipping away at the mortgage interest tax deduction*. Washington, DC: Brookings.

Goetz, E. G. (2013). *New deal ruins: Race, economic justice, and public housing policy*. Ithaca, NY: Cornell University Press.

Hays, D., & Sullivan, B. (2022). The wealth of households: 2022. *Current Population Reports*, P70BR-181. U.S. Census Bureau, Washington, DC.

Hoover, H. (1922). *American individualism*. New York: Doubleday, Page & Company.

Housing Trust Fund Project. (2021). *State housing trust fund revenues 2021*. Portland, OR: Community Change.

Jensen, L. S. (2015). Social provision before the twentieth century. In D. Beland, C. Howard, & K. J. Morgan (Eds), *The Oxford handbook of U.S. social policy*. Oxford: Oxford University Press.

Katz, M. B. (1996). *In the shadow of the poorhouse*. New York: Basic Books.

Keating, W. D. (2021). The right to housing: The goal versus the reality. In K. B. Anacker, M. T. Nguyen, & D. P. Vardy (Eds), *The routledge handbook of housing policy and planning* (pp. 9–22). New York: Routledge.

Kemeny, J. (1995). *From public housing to social market: Rental policy strategies in comparative perspective*. London: Routledge.

Krivo, L. J., & Kaufman, R. L. (2004). Housing and wealth inequality: Racial-ethnic differences in home equity in the United States. *Demography*, 41, 585–605.

Marcuse, P. (1995). Interpreting "public housing" history. *Journal of Architectural and Planning Research*, *12*(3), 240–258.

McCabe, B. J. (2016). *No place like home: Wealth, community & the politics of homeownership*. Oxford: Oxford University Press.

McClure, K. (2021). Subsidized rental housing programs. In K. B. Anacker, M. T. Nguyen, & D. P. Vardy (Eds), *The Routledge handbook of housing policy and planning*. New York: Routledge.

Miller, G. (2017). Maps show how tearing down city slums displaced thousands. *National Geographic*, December 15.

National Alliance of Community Economic Development Associations. (2014). *What is a Community Development Corporation?* Retrieved from www.naceda.org/index.php?option=com_dailyplanetblog&view=entry&category=bright-ideas&id=25%3Awhat-is-a-community-development-corporation-&Itemid=171

OECD Affordable Housing Database. Retrieved from www.oecd.org/els/family/PH4-2-Social-rental-housing-stock.pdf

O'Regan, K. M., & Quigley, J. M. (2000). Federal policy and the rise of nonprofit housing providers. *Journal of Housing Research*, *11*(2), 297–317.

Ramsey-Musolf, D. (2017). State mandates, housing elements, and low-income housing production. *Journal of Planning Literature*, *32*(2), 117–140.

Riis, J. (1890). *How the other half lives: Studies among the tenements of New York*. New York: Charles Scribner's Sons.

Roosevelt, F. D. (1937). FDR's Second Inaugural [Transcript]. Retrieved from https://teachinghistory.org/best-practices/examples-of-historical-thinking/25174

Rothstein, R. (2017). *The color of law: A forgotten history of how our government segregated America*. New York: Liveright Publishing Corporation.

Sanfilippo, A. (2018). *Who are the homeowner heroes?* National Association of Realtors. Retrieved from https://homeownershipmatters.realtor/issues/who-are-the-homeowner-heroes/#:~:text=After%20the%20First%20World%20War,aimed%20at%20encouraging%20home%20ownership.

Schwartz, A. (2014). *Housing policy in the United States*. New York: Routledge.

Stephens, M., & Fitzpatrick, S. (2007). Welfare regimes, housing systems and homelessness: How are they linked? *European Journal of Homelessness*, *1*(December), 201–212.

Urban Institute. (2020). *State and local backgrounders. Housing and community development expenditures*. Retrieved from www.urban.org/policy-centers/cross-center-initiatives/state-and-local-finance-initiative/state-and-local-backgrounders/housing-and-community-development-expenditures

Vale, L. (2007). *From the puritans to the projects: Public housing and public neighbors*. Cambridge, MA: Harvard University Press.

Williams, S. (2021). *The history of state housing finance agencies*. National Council of State Housing Agencies. Retrieved from https://statehfahistory.ncsha.org/wp-content/uploads/2021/09/NCSHA_DigitalBook_Final.pdf

Williamson, A. R. (2011). Can they afford the rent? Resident cost burden in Low Income Housing Tax Credit Developments. *Urban Affairs Review*, *47*, 775–799.

Chapter 5

Race and Affordable Housing

Quantifying Racial and Ethnic Inequities in Housing

Multiple disparities emerge when analyzing the impact of segregation and racism in the housing market. First, rates of homeownership are lower for households of color and this disparity has negative consequences for household wealth (Wachter & Acolin, 2022). Second, households of color face higher housing cost burdens (Hess et al., 2020). These differences can be attributed to lower incomes, but these gaps persist even after accounting for differences in socioeconomic status. Third, access to quality education is limited for households of color given racially segregated neighborhoods and the implications of that neighborhood structure on school quality (Owens, 2020; Reeves & Halikias, 2016). Finally, evictions and homelessness – two of the most significant (and negative) housing outcomes – are disproportionately experienced by Black, Hispanic, and Indigenous households and individuals (National Alliance to End Homelessness, 2023; Hepburn et al., 2020).

The accumulation of race-based disadvantages in the housing market is readily apparent. Households of color are more likely to be renters than homeowners – 59% of Black, 48% of American Indian/Alaska Native, 54% of Latino, and 41% of Asian households are renters compared to 28% of white non-Latino households (National Low Income Housing Coalition, 2022). Disparities in homeownership significantly impact the wealth gap since owning a home is a primary source of household wealth. As a consequence, significant race-based wealth gaps exist, and persist, in the United States. The median white household's net worth was nearly eight times the net worth of the median Black household, and over five times the median Latino household (National Low Income Housing Coalition, 2022). Unequal access to mortgage finance and racially restricted covenants have combined to reduce homeownership for households of color (Rothstein, 2018). Given the importance of homeownership in creating household wealth, these factors can be considered primary culprits for the significant race-based wealth gaps that we observe.

In addition to wealth gaps, there are pronounced racial disparities in household income. Lower incomes have an obvious negative impact on housing affordability. Households of color have persistently higher housing cost burdens compared to similarly situated white households (Hess et al., 2020; Joint Center for Housing Studies, 2022). In 2020, 55% of Black and 53% of Latino renters were cost-burdened, compared to 43% of white renters (National Low Income Housing Coalition, 2022). These racial differences persist even after controlling for socioeconomic status, household size, and household composition (Hess et al., 2020).

Given the strong relationship between home prices and educational quality, unequal access to quality housing and highly segregated neighborhoods can exacerbate race-based disparities in educational outcomes. Opportunity hoarding, which may keep low-income families out of neighborhoods with highly-ranked schools, may exacerbate disparities and maintain the status quo (Reeves & Halikias, 2016). Such patterns reinforce high house prices near the best schools and less expensive housing near lower performing schools (Reeves & Halikias, 2016). Across the 100 largest metro areas in America, housing costs are on average 2.4 times higher near better performing public schools compared to housing near lower performing public schools (Rothwell, 2012). The consequences of these dynamics are twofold. First, home prices near good schools increase faster, thus increasing the household wealth of homeowners (frequently white homeowners) who live near those schools. Second, if you believe that quality schools are a tool for upward social mobility, highly segregated housing may limit or inhibit upward mobility of children who attend lower-ranked schools (frequently children of color).

An obvious consequence of racial disparities in the housing system are the profound differences in the prevalence of evictions and homelessness. Black households are more likely to be evicted, with an average eviction filing rate in 2020 of 6.2% and an eviction rate of 3.4%, compared to an average eviction filing rate of 3.4% and eviction rate of 2.0% for white renters (Hepburn, Louis, & Desmond, 2020). Rates of eviction are noticeably higher for Black and Latina women renters compared to Black and Latino males (Hepburn et al., 2020). The consequences of eviction are well-documented. For example, eviction is often a precipitating cause of homelessness (Collinson, 2023). In 2022, even though white individuals represent half of the homeless population, people of color experience higher rates of homelessness. Eleven out of every 10,000 white people, 48 out of every 10,000 Black people, and 121 out of every 10,000 Native Hawaiian or Pacific Islander people experience homelessness (National Alliance to End Homelessness, 2023).

As the evidence demonstrates, households from different racial and ethnic backgrounds face significantly lower homeownership rates and wealth accumulation, higher housing cost burdens, and are more likely to experience eviction and homelessness than other similarly situated white households. The remaining sections of this chapter examine how residential segregation and discrimination in the housing market drives racial disparities and the outcomes linked to these disparities.

Residential Segregation

Residential segregation divides groups of people into different geographic spaces. This division may occur based on income, race, ethnicity, religion, or cultural differences. The origins of residential segregation can be traced back to the early history of the U.S.

There are three theories that seek to explain the conspicuous segregation that is evident across the United States. Darden provides a summary of these theories in his chapter in the *Encyclopedia of Housing* (Darden, 2012). The first theoretical explanation for segregation is *class theory*. This theory suggests that people are segregated because of differences in socioeconomic status (income, education, and occupational status). This logic suggests that if Black households have a lower standing in society – from a socioeconomic standpoint – they will reside in less expensive housing. These class differences, therefore, sort households into different neighborhoods thus producing segregated communities. The second theory is called *voluntary segregation* or *preference theory*. This explanation argues that people prefer to live in close proximity to other people of a similar race or ethnicity. As a result – according to the theory – Black households tend to cluster in the same neighborhoods and produce segregation. Finally, there is the *racial discrimination* or *racial steering theory*. This theory posits that discriminatory and racist behavior by banks, insurance companies, policymakers, landlords, and real estate agents have limited access to housing for households of color, thus producing the significant segregation that currently exists. Darden (2012) suggests that the balance of the evidence supports this last theory as providing the most credible explanation for segregation.

Sociologists use the theories of place stratification and spatial assimilation to explain residential segregation. Place stratification focuses on the economic and social factors that contribute to segregation, including unequal access to resources like education. This leads to certain groups being sorted into specific areas of a community, further widening existing inequalities. Spatial assimilation, on the other hand, describes how minority groups enter neighborhoods with more resources in order to integrate and move up socioeconomically. While place stratification highlights the perpetuation of segregation, spatial assimilation explains the process of integration within these segregated environments. Together, these theories help us understand the complex dynamics of residential segregation.

Numerous decisions, policies, and actions over time support the idea that racial discrimination has been a primary cause of residential segregation. Zoning and land use have had a significant impact on segregation and the racial composition of cities. After racial segregation through zoning was ruled unconstitutional in 1917 (*Buchanan v. Warley, 245 U.S. 60*, 1917), cities sought to maintain existing patterns of segregation through land use and zoning, creating more restrictive zoning in predominantly white areas at the outskirts of cities. This allowed communities to prevent industrial buildings and dense housing from being built near those neighborhoods. On the other hand, in predominantly Black areas, zoning regulations were less restrictive, exposing minorities to a disproportionate level of locally unwanted land uses, also referred to as LULUs, (Shertzer et al., 2022). Zoning that restricted housing density and inexpensive housing in affluent neighborhoods is associated with greater residential segregation (Rothwell & Massey, 2010; Rothwell, 2012). Cities with more restrictive land use tended to remain whiter from 1970 to 2011, while cities whose zoning regulations were challenged in court, and were threatened or forced by courts to liberalize their land use laws saw a growth in households of color from 1970 to 2011 (Trounstine, 2020).

In addition to land use, many other actions have contributed to the segregation we observe. Outright discrimination by landlords and real estate agents have prevented households of color from renting and purchasing housing in predominantly white neighborhoods (Galster, 1988). Furthermore, the availability of mortgage financing was constrained by a host of policies and procedures, notably redlining. The effect of these policies was to limit homeownership opportunities for households of color and to restrict residential mobility, thus perpetuating segregation. Even low-income housing policy has contributed to segregation. The decisions made by local authorities regarding the siting of affordable housing has served to concentrate recipients of housing assistance (many of whom are households of color) in highly segregated neighborhoods. In this way, federal housing policies have exacerbated, rather than mitigated, racial segregation in housing. Finally, race-based income disparities that are well-documented limit the purchasing power of households of color (National Low Income Housing Coalition, 2022). Given the constrained purchasing power, households of color may struggle to afford the higher housing costs that are frequently found in predominantly white neighborhoods.

Recent research sheds new light on the persistent nature of residential segregation in the United States. Years after passage of the Fair Housing Act, levels of segregation have not fallen meaningfully. This is somewhat surprising because incomes have risen for households of color and there has been more aggressive regulation to prevent discrimination. Why might this be the case? Maria Krysan and Kyle Crowder, in their book, *Cycle of Segregation: Social Processes and Residential Stratification*, show how various social processes continue to contribute to highly segregated neighborhoods. This includes daily activities, social networks, and lived experiences that lead to blind spots during the housing search process that reinforce the cycle of residential stratification. Because individuals tend to depend on social networks that are often homogenous and gravitate toward what is familiar, these factors play an important role in determining where one might live in the future. Therefore, these social processes may prevent or slow reductions in segregation despite macro factors that might suggest otherwise (Krysan & Crowder, 2017). In this argument, we see evidence of two theories described above: the racial discrimination theory *and* the preference theory.

Measuring Segregation

There are five primary approaches used to measure residential segregation, including evenness, exposure, concentration, centralization, and clustering (Massey & Denton, 1988). Within these broad categories, over twenty different segregation indices have been created. One of the original and most common indices used is the index of dissimilarity. This index was introduced in the 1950s after several decades of debate and was prompted by high levels of racial segregation at that time (Duncan & Duncan, 1955; Burgess, 1928). The index of dissimilarity measures the inequality in the distribution of one group in a space relative to another group. The metric is used to compare the distribution of two sub-groups to determine the proportion of the population that would have to relocate to obtain an even geographic distribution (each group has the same percentage of members in all neighborhoods as represented overall in the metropolitan region).

Using the index of dissimilarity, segregation peaked in 1970, when 80% of the Black population would have had to move to eliminate racial segregation (Glaeser & Vigdor, 2012). Segregation has been decreasing ever since; by 2010 approximately 55% of the Black population would have to move to eliminate segregation. Despite improvements, levels of segregation remain high and in certain markets, such as Boise (Idaho), Portland (Maine), and Manchester (New Hampshire), segregation has continued to increase (Glaeser & Vigdor, 2012). Furthermore, different races and ethnicities continue to live in very different neighborhoods. White households remain living in predominately white neighborhoods whereas non-whites are more concentrated in more ethnically-diverse neighborhoods (Logan & Stults, 2021).

The entropy index is like the dissimilarity index but can account for more than two groups and is often referred to as a multiethnic measure of segregation (Reardon & Firebaugh, 2021). Other similar metrics that measure evenness of a population are the Gini coefficient and Atkinson index, which are often used as indicators of economic inequality instead of racial segregation. Additional indices of residential segregation have been created to capture the other four dimensions of residential segregation: exposure, the degree of potential contact or possibility of interaction; clustering, the extent to which areas with subpopulations adjoin one another; centralization, the degree to which a group is located near an urban center; and concentration, the amount of physical space occupied by a subpopulation (Massey & Denton, 1988). When a subpopulation experiences extreme segregation on all dimensions this is known as hypersegregation. Research demonstrates that Black individuals, more than any other group, have experienced multidimensional hypersegregation, highlighting the impact of social isolation in understanding wealth disparities (Massey & Denton, 1989).

Historically, measures of segregation have treated each spatial unit (e.g., each neighborhood) independent of those around it (e.g., adjacent neighborhoods) and have been limited by their inability to account for spatial autocorrelation, that is the tendency for areas near each other to be similar. More recently, measures have been developed to incorporate spatial relationships across an area (Yao et al., 2018). As one example, the Spatial Gini Index of segregation can be used to estimate the spatial arrangement of racial groups by quantifying the degree of overall segregation from racial clustering of adjacent neighborhoods. Applying this to Atlanta, it was found that racial segregation ratios differ depending on proximity to the central business district, with neighborhoods near the city center more likely to be segregated, and neighborhoods further from the city center more likely to be integrated (Dawkins, 2004). Another example is the Index of Concentration at the Extremes (ICE), which is a measure that examines the spatial polarization of racial and social groups. The ICE identifies racial and/or economic households that are "most deprived" and "most privileged." This index was applied to preterm birth and infant mortality in the state of California, and it was found that Black women were most likely to live in extreme race and income concentrations, and these areas increased the odds of preterm birth and infant mortality (Chambers et al., 2019).

Frequently, segregation is focused on the concentration of poor households of color. Racially/Ethnically Concentrated Areas of Poverty is an official focus of policy by the U.S. Department of Housing and Urban Development. A study from researchers at the University of Minnesota took a different approach to understanding segregation. They flipped the script and instead focused on concentrations of affluent white households in certain areas. They named these areas, Racially Concentrated Areas of Affluence (Goetz et al., 2019). The findings of this study demonstrate that it is not just poor households of color that concentrate in certain neighborhoods, affluent white households do the same. The authors of the study highlight the need to focus on policies – such as exclusionary land use and the home mortgage interest tax deduction – that have helped to produce these unequal communities (Goetz et al., 2019).

Discrimination in the Housing Market

As described briefly in the prior sections, multiple forms of explicit racial discrimination occur in the housing market, including unequal access to information, steering, and blockbusting (Galster, 1988). In addition, implicit forms of racial discrimination further limit access to housing for lower income households and households of color. Land use policies, for example, can serve to exacerbate racial segregation and limit access to resource-rich neighborhoods. Single-family zoning policies allow many local jurisdictions to restrict the development of multifamily housing to certain areas in a city. Zoning may also exclude smaller homes by regulating minimum lot sizes and homes, which limits affordable housing production. Affordable housing is often subject to Not In My Backyard (NIMBY) opposition since development is perceived to have negative impacts on a local community due to unfounded fear about the type of people who live in affordable housing. This section explores a host of discriminatory practices found in the housing market that perpetuate residential segregation.

Exclusionary Practices and Policies in Homeownership

There are a range of discriminatory practices – in the real estate and banking industries – that have influenced the locational outcomes of households of color and have produced and exacerbated residential segregation. Three primary forms of location-based discrimination in the real estate industry include deed restrictions, blockbusting, and steering. While zoning based on race was deemed unconstitutional in 1917 (*Buchanan v. Warley, 245 U.S. 60*, 1917), racially restrictive covenants, known as *deed restrictions*, were ruled constitutional in 1926 (*Corrigan v. Buckley, 271 U.S. 323*, 1926) and began to be widely used throughout the United States. These deeds prevented Black families (and other racial or ethnic minority households) from purchasing homes in white neighborhoods. In addition to the legally restrictive deeds, there was an additional layer of social enforcement as white residents threatened Black households with violence if they attempted to purchase a home in their neighborhood. Real estate brokers received similar threats if they attempted to sell a home to a Black household in a white neighborhood.

A second form of discrimination that promoted segregation in homeownership was *blockbusting*. This strategy was employed by housing speculators to prey on the insecurities and discriminatory impulses of white households. Blockbusters would approach white homeowners in neighborhoods that were adjacent to predominantly Black neighborhoods and encourage them to sell their homes at discounted prices to speculators. The implicit threat was that the close proximity to a Black neighborhood might reduce the value of homes owned by white households in the near vicinity, therefore motivating white households to sell their homes at a discount. The homes were then sold at an inflated price to Black families who had limited options for homeownership because of multiple forms of racism and discrimination. By 1962, blockbusters in Chicago changed two to three blocks a week from predominantly white to predominantly Black. In 1968, blockbusting largely ended after successful litigation deemed the practice to be illegal (Mehlhorn, 1998).

A third practice that produced residential segregation in the homeownership market is *steering*, in which real estate brokers steer prospective homebuyers to homogeneous neighborhoods that match their race or ethnicity. A 1960s study conducted in Chicago found that a majority of real estate brokers supported exclusionary practices (Helper, 1969). While prevalent throughout the second half of the twentieth century, steering continues to be practiced today. Interviews with New York real estate agents in 2017 revealed that agents continue to steer or guide buyers toward neighborhoods that align with their race or ethnicity and to solicit buyers for homes that "fit with the neighborhood" (Besbris & Faber, 2017). Sociologist Elizabeth Korver-Glenn argues housing market professionals perpetuate segregation through racialized routines in the normal course of business using ethnographic and interview data in Houston, TX (Korver-Glenn, 2021). A frequently delivered defense of this behavior is that Black households prefer to live in all Black neighborhoods. But existing research demonstrates that this is not necessarily the case. Although some Black households prefer to remain in Black neighborhoods (Krysan & Crowder, 2017), interviews with 2,000 Black residents in the mid-1990s found that a majority preferred greater integration and would move into a white neighborhood were it not for fear of hostility from incumbent white residents (Krysan & Farley, 2002). Furthermore, research indicates real estate agents steer Black households away from white neighborhoods, specifically Black householder females with children (Hall et al., 2023).

The banking and mortgage finance industry also had discriminatory policies and procedures that complemented – and exacerbated – the practices of the real estate industry. Combined, these institutional approaches were powerful forces for segregation. At a basic level, decades of discrimination in mortgage lending limited access to homeownership for households of color and perpetuated residential segregation. Historically, the Federal Housing Administration (FHA) insured loans for newly constructed homes that primarily served white households in suburban neighborhoods. The FHA enacted discriminatory lending policies, known as redlining, in which they conducted housing market analyses, developed a risk rating system, and created an extensive collection of maps on a block-by-block basis which dictated where to insure mortgages; such insurance was typically provided to homes in predominantly white neighborhoods

(Hillier, 2003). From 1934 to 1962, only 2% of the $120 billion in FHA insured loans were provided to nonwhite families (Solomon et al., 2019). Redlining has helped to produce highly segregated neighborhoods and has had lasting impacts on health, social and economic outcomes, and homeownership rates. While redlining plays an important role in the housing history of the U.S., the impacts are still felt today. Formerly redlined neighborhoods still have a higher likelihood of lending discrimination measured by lower loan volume and higher interest rates (Lynch et al., 2021) and white middle-income neighborhoods are still more likely to receive loans than Black middle-income neighborhoods (Wyly & Holloway, 1999).

The GI Bill, established in 1944, has helped millions of military veterans purchase a home. The bill is frequently credited as one of the drivers of prosperity for middle-income households that was evident in the decades following World War II. What is less understood is how the GI Bill disproportionately benefited white families, by allowing local banks to discriminate against Black veterans and deny them home loans. For example, in Mississippi, where half of the population is Black, just two of the 3,000 mortgages that the VA guaranteed in 1947 went to Black households (Solomon et al., 2019). Such disproportionate treatment was, in essence, baked into the law because great discretion and latitude was provided to individual states. Therefore, those states with Jim Crow laws implemented the GI Bill in ways that prevented potential Black homeowners from benefitting. In certain states, the law was implemented in ways in which Black veterans received unreliable information on how to access benefits from the GI Bill or were discouraged from doing so. Furthermore, GI benefits were limited to veterans who received an honorable discharge. Because Black people in the military were disproportionately court-marshalled or unjustly dishonorably discharged (Connecticut Veterans Legal Center, 2022), they were not eligible to benefit from the GI Bill.

Finally, predatory lending occurs when lenders obscure information from borrowers, pressure them into loans with which they are uncomfortable, or knowingly provide mortgages to individuals who cannot afford them. Based on data analyzed from the 2004 Home Mortgage Disclosure Act database, Black people were less likely to receive loans from regulated lenders that provide the best loan terms compared to white Americans and were more likely to receive high priced, predatory loans (Beeman et al., 2010). Real estate transaction costs for low-income households also impact affordability. There are substantial differences in loan and closing cost fees for households of color and buyers with lower levels of educational attainment (Apgar & Calder, 2005; Woodward, 2003; Woodward, 2008). These actions can all increase the cost of homeownership, thus decreasing opportunity, restricting wealth accumulation, and increasing the risk of foreclosure for households that receive disproportionately higher cost mortgages.

Exclusionary Practices and Policies in Rental Markets

Discriminatory practices are not only found in the homeownership market. We observe significant race-based impediments in the market for rental housing as well. These exclusionary practices and policies in the rental market involve both overt and indirect discrimination. Rental advertisements differ for white versus non-white neighborhoods and display racialized neighborhood discourse – although explicit racial language has been outlawed since 1968 (Kennedy et al., 2021). Segregation is also perpetuated by online rental advertisements in which predominantly white neighborhoods are over-represented (Boeing, 2020; Hess et al., 2021). Rental listings on Craigslist, an online rental platform, largely represent neighborhoods that are college-educated, English-only speaking populations aged 20–34, while Black and Hispanic neighborhoods are underrepresented (Boeing, 2020). In cities with greater levels of Black-white segregation, majority Black neighborhoods tend to be underrepresented on Craigslist rental ads (Hess et al., 2021). Online rental housing search platforms that are designed for households with housing assistance, such as GoSection8, have more units advertised in higher poverty neighborhoods compared to Craigslist, Apartments.com, and Zillow (Hess et al., 2023).

Exclusionary practices in rental markets go beyond rental advertisements, which is evident when examining where affordable housing is built. NIMBYism has kept affordable housing out of certain neighborhoods and impacted where developers build affordable rental housing. A survey of New York City developers found that 70% of them experienced local opposition to affordable housing developments, causing delays, forcing changes, increasing costs, and in some cases pushing the developments into lower-income areas where there is less political strength to oppose development (Scally & Tighe, 2015). Similar patterns emerge when siting Low Income Housing Tax Credit (LIHTC) developments. Opposition and NIMBYism emerges which causes developers to identify the "path of least resistance" which means constructing new affordable housing in neighborhoods with higher concentrations of poverty and households of color (Dawkins, 2004). Evidence suggests that all affordable housing development, not just units constructed pursuant to the LIHTC program, face similar obstacles and challenges.

In response to the historical siting of affordable housing, and in light of ongoing challenges associated with the development of affordable housing, programs like the Housing Choice Voucher (HCV) have been established to provide housing "choice" to households in the private market. However, very few households with vouchers move to low-poverty neighborhoods, and low-income households are often kept out of higher-income neighborhoods (Cunningham et al., 2018; Tighe et al., 2017). Discrimination toward low-income households with housing assistance continues to exacerbate patterns of residential segregation. Cunningham and colleagues (2018) found that many landlords reject households with vouchers but the denial rate varies by place. Fort Worth (78%), Los Angeles (76%), and Philadelphia (67%) have some of the highest denial rates from landlords (Cunningham et al., 2018).

Neighborhood Effects

Discrimination in housing markets have perpetuated residential segregation and, as a result, households of color are overly represented in under-resourced neighborhoods. There are a host of negative consequences with residing in such neighborhoods. Low-income and low-cost neighborhoods often have higher levels of crime, lower-quality schools, limited access to public transit, are less walkable, and have less access to amenities and essential services (Gabriel & Painter, 2020; Swope & Hernández, 2019). These results underscore why high levels of residential segregation are problematic and why access to affordable housing in a range of different neighborhoods is essential.

Under-resourced neighborhoods experience higher crime rates and have been linked to higher rates of gun violence (Stansfield & Semenza, 2021). This can isolate families by pushing more activities into the home to avoid potentially dangerous outdoor environments (Hernández, 2016). More time spent indoors can lead to social isolation and a range of poor mental and physical health outcomes. For example, areas with higher crime are associated with higher asthma rates (Beck et al., 2016). Living in dangerous neighborhoods also takes a toll on mental health and wellbeing. Stress and anxiety increase in unsafe environments and these factors have been linked to a host of poor health outcomes including depression and heart disease (Eberly et al., 2022; Weisburd et al., 2018).

Housing units that are more affordable are often located in under-performing school districts. Children from low-income households attend schools that rank meaningfully lower than the schools that children from moderate- and high-income households attend (Turner & Gourevitch, 2017). Because of the strong link between education, college attendance, and earnings in adulthood, the relationship between educational access and affordability underscores how segregated neighborhoods can perpetuate racial wealth gaps.

Under-resourced neighborhoods also have a disproportionate presence of environmental hazards. Carcinogenic pollutants, poor air quality, proximity to highways and traffic, and undesirable land uses such as landfills are more prominent in areas with affordable housing. Research indicates that lower-income neighborhoods have greater exposure to air pollutants compared to white households in higher income neighborhoods (Bell & Ebisu, 2012). The most significant environmental disparities are observed in metropolitan areas with higher levels of residential segregation (Woo et al., 2019).

There is a notable lack of neighborhood amenities in under-resourced and highly segregated neighborhoods. Access to essential public services such as public transit, healthy food, recreational facilities, and early childhood development programs are often absent from lower-income, highly segregated neighborhoods. Neighborhoods that promote physical activity, such as greenspace for children to play and exercise, and sidewalks for walkability are more prevalent in higher-income and disproportionately white areas (Rigolon, 2016; Watson et al., 2016). Low-income neighborhoods also host high levels of noise pollution, inadequate lighting, pest infestations, extreme temperature changes, and safety issues that can be a source of anxiety and impact mental health (Hoisington et al., 2019). In a cruel twist, while residents of low-income neighborhoods face a range of threats to physical and mental wellbeing, access to quality healthcare is also limited (Prentice, 2006). This brief summary of neighborhood effects highlights how location and context can have an adverse impact on the experiences of local residents.

Complicating the neighborhood effects analysis is the issue of gentrification. Investment into previously disinvested neighborhoods has many benefits. But, as research has demonstrated, such investments can lead to the gentrification of a neighborhood which can contribute to its own set of negative outcomes, including displacement. Hwaing and Ding (2020) demonstrate that the negative displacement effects are disproportionately affecting households of color. While gentrification can create benefits for some incumbent residents, Hwang and Ding (2020) highlight that the negative costs of gentrification are most pronounced for low-income residents of neighborhoods in gentrifying neighborhoods that were previously predominantly Black. Therefore, it is important for readers to understand that increasing investment alone, may not be a sufficient response to the negative neighborhood effects outlined in this section. Rather, there are a range of policy tools that communities can use to mitigate the consequences of gentrifying neighborhoods, including capping property taxes for incumbent low and middle-income homeowners. Increasing the resource base in non-gentrifying neighborhoods is also important because it provides greater opportunities for positive moves for households that choose (or are forced) to exit gentrifying neighborhoods.

Conclusion

In this chapter, we seek to highlight the important relationship between discrimination, segregation, and affordable housing. Where people live really matters. Therefore, to construct an equitable and affordable residential built environment, we must create quality, affordable housing in all neighborhoods. And we must ensure that the tools we use to provide affordability do not exacerbate prior patterns of segregation and discrimination. In sum, we must provide avenues for residential mobility for those who choose to move *and* we must invest in communities that have been sites of historical disinvestment for those who choose to remain in their home communities. Failure to do so will allow harmful historical trends to persist.

References

Apgar, W. C., & Calder, A. (2005). *The dual mortgage market: The persistence of discrimination in mortgage lending*. Cambridge, MA: Harvard University, Joint Center for Housing Studies.

Beck, A. F., Huang, B., Ryan, P. H., Sandel, M. T., Chen, C., & Kahn, R. S. (2016). Areas with high rates of police-reported violent crime have higher rates of childhood asthma morbidity. *The Journal of Pediatrics, 173*, 175–182.

Beeman, A., Silfen, D., Glasberg, S., & Casey, C. (2010). Whiteness as property: Predatory lending and the reproduction of racialized inequality. *Critical Sociology, 31*(1), 27–25.

Bell, M. L., & Ebisu, K. (2012). Environmental inequality in exposures to airborne particulate matter components in the United States. *Environmental Health Perspectives, 120*(12), 1699–1704.

Besbris, M., & Faber, J. W. (2017). Investigating the relationship between real estate agents, segregation, and house prices: Steering and upselling in New York state. *Sociological Forum, 32*(4), 850–873.

Boeing, G. (2020). Online rental housing market representation and the digital reproduction of urban inequality. *EPA: Economy and Space, 52*(2), 449–468.

Burgess, E. W. (1928). Residential segregation in American Cities. *Annals of the American Academy of Political and Social Science, 14*, 105–115.

Chambers, B. D., Baer, R. J., McLemore, M. R., & Jelliffe-Pawlowski, L. L. (2019). Using Index of Concentration at the Extremes as indicators of structural racism to evaluate the association with preterm birth and infant mortality – California, 2011–2012. *Journal of Urban Health, 96*, 159–170.

Collinson, R., Humphries, J. E., Mader, N. S., Reed, D. K., Tannenbaum, D. I., & van Dijk, W. (2023). *Eviction and poverty in American cities*. NBER Working Paper Series 30382.

Connecticut Veterans Legal Center. (2022). *How Racial Disparities in the Military's Administrative Separation System Harm Black Veterans*. Retrieved from http://ctveteranslegal.org/wp-content/uploads/2022/11/Discretionary-Injustice-Report.pdf

Cunningham, M. K., Galvez, M. M., Aranda, C., Santos, R., Wissoker, D., Oneto, A. D., Pitingolo, R., & Crawford, J. (2018). *A pilot study of landlord acceptance of housing choice vouchers*. Washington, DC: Urban Institute.

Darden, J. T. (2012). "Segregation." In A. T. Carswell (Ed.), *The Encyclopedia of Housing*, 2nd edition. Thousand Oaks, CA: Sage Publishing.

Dawkins, C. J. (2004). Measuring the spatial pattern of residential segregation. *Urban Studies, 41*(4), 833–851.

Duncan, O. D., & Duncan, B. (1955). A methodological analysis of segregation indices. *American Sociological Review, 20*, 210–217.

Eberly, L. A., Julien, H., South, E. C., Venkataraman, A., Nathan, A. S., Anyawu, E. C., Dayoub, E., Groeneveld, P. W., & Khatana, S. A. M. (2022). Association between community-level violent crime and cardiovascular mortality in Chicago: A longitudinal analysis. *Journal of the American Heart Association, 11*(14), e025168.

Gabriel, S., & Painter, G. (2020). Why affordability matters. *Regional Science and Urban Economics, 80*, 103378.

Galster, G. (1988). Residential segregation in America cities: A contrary review. *Population Research and Policy Review, 7*, 93–112.

Glaeser, E., & Vigdor, J. (2012). *The end of the segregated century: Racial separation in America's neighborhoods, 1890–2010*. New York, NY: Manhattan Institute for Policy Research, Center for State and Local Leadership.

Goetz, E., Damiano, A., & Williams, R. (2019). Racially concentrated areas of affluence. *Cityscape, 21*(1), 99–124.

Hall, M., Timberlake, J.M., & Johns-Wolfe, E. (2023). Racial steering in U.S. housing markets: When, where, and to whom does it occur? *Social Research for a Dynamic World, 9*, 1–17.

Helper, R. (1969). *Racial policies & practices of real estate brokers*. Minneapolis, MN: University of Minnesota Press.

Hepburn, P., Louis, R., & Desmond, M. (2020). Racial and gender disparities among evicted Americans. *Sociological Science*, 7, 649–662.

Hess, C., Acolin, A., Walter, R., Kennedy, I., Chasins, S., & Crowder, K. (2021). Searching for housing in the digital age. Neighborhood representation on internet rental housing platforms across space, platform, and metropolitan segregation. *EPA: Economy and Space*, 53(8), 2012–2032.

Hess, C., Colburn, G., Crowder, K., & Allen, R. (2020). Racial disparity in exposure to housing cost burden in the United States: 1980–2017. *Housing Studies*, 37(10), 1821–1841.

Hess, C., Walter, R. J., Kennedy, I., Acolin, A., Ramiller, A., & Crowder, K. (2023). Segmented information, segregated outcomes: Housing affordability and neighborhood representation on a voucher-focused online housing platform and three mainstream alternatives. *Housing Policy Debate*, 33(6), 1511–1535.

Hillier, A. E. (2003). Redlining and the Homeowners' Loan Corporation. *Journal of Urban History*, 29(4), 394–420.

Hoisington, A. J., Stearns-Yoder, K. A., Schuldt, S. J., Beemer, C. J., Maestre, J. P., Kinney, K. A., Postolache, T. T., Lowry, C. A., & Brenner, L. A. (2019). Ten questions concerning the built environment and mental health. *Building and Environment*, 155, 58–69.

Hwang, J., & Ding, L. (2020). Unequal displacement: Gentrification, racial stratification, and residential destinations in Philadelphia. *American Journal of Sociology*, 126(2), 354–406.

Joint Center for Housing Studies. (2022). *The state of the nation's housing*. Cambridge, MA: Harvard University.

Kennedy, I., Hess, C., Paullada, A., & Chasins, S. (2021). Racialized discourse in Seattle rental ad texts. *Social Forces*, 99(4), 1432–1456.

Korver-Glenn, E. (2021). *Race Brokers: Housing markets and segregation in 21st century urban America*. Oxford University Press.

Krysan, M., & Crowder, K. (2017). *Cycle of segregation: Social processes and residential stratification*. New York: Russell Sage Foundation.

Krysan, M., & Farley, R. (2002). The residential preferences of Blacks: Do they explain persistent segregation? *Social Forces*, 80(3), 937–980.

Logan, J., & Stults, B. (2021). *The persistence of segregation in the metropolis: New findings from the 2020 census*. Diversity and disparities project, Brown University. Retrieved from https://s4.ad.brown.edu/Projects/Diversity/Data/Report/report08122021.pdf

Lynch, E. E., Malcoe, L. H., Laurent, S. E., Richardson, J., Mitchell, B. C., & Meier, H. C. S. (2021). The legacy of structural racism: Associations between historic redlining, current mortgage lending, and health. *Population Health*, 14, 1–10.

Massey, D. S., & Denton, N. A. (1988). The dimensions of residential segregation. *Social Forces*, 67(2), 281–315.

Massey, D. S., & Denton, N. A. (1989). Hypersegregation in U.S. metropolitan areas: Black and Hispanic segregation along five dimensions. *Demography*, 26(3), 373–391.

Mehlhorn, D. (1998). A requiem for blockbusting: Law, economics, and race-based real estate speculation. *Fordham Law Review*, 67(3), 1145–1192.

National Alliance to End Homelessness. (2023). *State of homelessness: 2023 edition*. Retrieved from https://endhomelessness.org/homelessness-in-america/homelessness-statistics/state-of-homelessness/#key-facts

National Low Income Housing Coalition (2022). *The gap. A shortage of affordable homes*. Washington, DC: National Low Income Housing Coalition.

Owens, A. (2020). Unequal opportunity: School and neighborhood segregation in the USA. *Race and Social Problems*, 12, 29–41.

Prentice, J. C. (2006). Neighborhood effects on primary care access in Los Angeles. *Social Science & Medicine*, 62(5), 1291–1303.

Reardon, S.F., & Firebaugh, G. (2002). Measures of multigroup segregation. *Sociological Methodology*, 32(1), 33–67.

Reeves, R. V., & Halikias, D. (2016). *How land use regulations are zoning out low-income families*. Washington, DC: Brookings Metropolitan Policy Program.

Rigolon, A. (2016). A complex landscape of inequity in access to urban parks: A literature review. *Landscape and Urban Planning*, 153, 160–169.

Rothstein, R. (2018). *The color of law: A forgotten history of how our government segregated America*. Liveright Publishing Corporation.

Rothwell, J. T. (2012). *Housing costs, zoning, and access to high-scoring schools*. Washington, DC: Brookings Metropolitan Policy Program.

Rothwell, J. T., & Massey, D. S. (2010). Density zoning and class segregation in U.S. metropolitan areas. *Social Science Quarterly, 91*(5), 1123–1143.

Scally, C., & Tighe, J. R. (2015). Democracy in action? NIMBY as impediment to equitable affordable housing siting. *Housing Studies, 30*(5), 749–769.

Shertzer, A., Twinam, T., & Walsh, R. P. (2022). Zoning and segregation in urban economic history. *Regional Science and Urban Economics, 94*, 1–8.

Solomon, D., Maxwell, C., & Castro, A. (2019). *Systematic inequality: Displacement, exclusion, and segregation*. Washington, DC: Center for American Progress.

Stansfield, R., & Semenza, D. (2021). Urban housing affordability, economic disadvantage and racial disparities in gun violence: A neighbourhood analysis in four US cities. *The British Journal of Criminology*, 1–19.

Swope, C. B., & Hernandez, D. (2019). Housing as a determinant of health equity: A conceptual model. *Social Science Medicine, 243*, 112571.

Trounstine, J. (2020). The geography of inequality: How land use regulation procures segregation. *American Political Science Review, 114*(2), 443–455.

Turner, M. A., & Gourevitch, R. (2017). *How neighborhoods affect the social and economic mobility of their residents*. Washington, DC: Urban Institute.

Wachter, S., & Acolin, A. (2022). Homeownership for the long run. *Journal of Comparative Urban Law and Policy, 5*(1), Article 23, 274–296.

Watson, K. B., Harris, C. D., Carlson, S. A., Dorn, J. M., & Fulton, J. E. (2016). Disparities in adolescents' residence in neighborhoods supportive of physical activity – United States, 2011–2012. *Morbidity and Mortality Weekly Report, 65*(23), 598–601.

Weisburd, D., Cave, B., Nelson, M., White, C., Haviland, A., Ready, J., Lawton, B., & Sikkema, K. (2018). Mean streets and mental health: Depression and post-traumatic stress disorder at crime hot spots. *American Journal of Community Psychology, 61*(3–4), 285–295.

Woodward, S. E. (2003). *Consumer confusion in the mortgage market*. Palo Alto, CA: Sand Hill Econometrics.

Woodward, S. E. (2008). *A study of closing costs for FHA mortgages*. Washington, DC: U.S. Department of Housing and Urban Development.

Wyly, E. K., & Holloway, S. R. (1999). 'The color of money' revisited: Racial lending patterns in Atlanta's neighborhoods. *Housing Policy Debate, 10*(3), 555–600.

Yao, J., Wong, D. W. S, Bailey, N., & Minton, J. (2018). Spatial segregation measures: A methodological review. *Economische en Sociale Geografie, 110*(3), 235–250.

Chapter 6

Housing Instability

Overview

Over the past 20 years, foreclosure, eviction, and homelessness have been exacerbated by a lack of affordable housing. Because the vast majority of housing in the United States is procured in the private market, the actions of lenders and landlords play a prominent role in the stability – or lack thereof – of housing for low-income households. Since 2000, there has been a credit-fueled housing boom, the subsequent housing market bust which contributed to the global financial crisis, growing demand for rental housing in the aftermath of the subprime mortgage crisis, an intensifying eviction crisis, and a growing recognition of the problem of homelessness across the United States. Housing – or more specifically housing affordability – has been at the center of many important social and economic events of the last 25 years. In this chapter, we summarize the importance of foreclosure, eviction, and homelessness in relation to affordable housing.

Foreclosure

Housing markets were center stage throughout the first decade of the 2000s. The first seven years of the decade produced house price appreciation unlike anything seen in the history of the United States. The average price of a residential home nearly doubled from 2000 to 2006 (U.S. Census Bureau, 2023) as lax lending standards and abundant mortgage finance provided historically easy access to homeownership opportunities. Easy credit, capitalistic motives, and supportive public policy drove home values to dizzying heights before an abrupt collapse in 2007. In addition to a substantial decline in home values – a reduction of over 25% from 2006 to 2010 according to the S&P/Case-Shiller U.S. National Home Price Index – the precipitous drop disrupted the entire global financial system as trillions of mortgage-backed securities (and insurance on those securities) lost most, if not all, of their value in a very short time. The financial system, and many titans of the industry – including Bear Stearns, Lehman Brothers, and AIG – were brought to their knees. Lost in the headlines at the time was the plight of six million individual homeowners who lost their homes in the foreclosure crisis (Makridis & Ohlrogge, 2021).

Foreclosure is a legal process where a bank, or lender, seeks to take control of an asset (in this case a home) when the borrower has stopped making payments. Most people who buy their homes do so with the help of mortgage financing. A mortgage, or loan, allows people to purchase homes that they otherwise would not be able to afford. In exchange for the money, the borrower agrees to make monthly payments on the loan until it is paid back. Conventional mortgages have thirty-year terms, so borrowers make 360 payments (12 payments a year over 30 years) until the loan has been fully paid off. One aspect of a mortgage loan is a security interest. The lender receives a security interest in the underlying asset – in this case the home of the borrower – and can take control of that home in the case of non-payment by the borrower. This security interest provides protection for the bank or lender in the event of non-payment. Once a bank gains control of the home through the foreclosure process, they can then do what they choose with the home. Typically, they will attempt to sell the home and use the proceeds from the sale of that home to cover all or part of the outstanding, and unpaid, debt.

One of the major triggers of the 2008–2010 subprime mortgage crisis was a loan product called an adjustable-rate mortgage (ARM). These types of mortgages have fixed interest rates for a set period of time – say five years – after which time those interest rates adjust based on the lender's prime rate. To entice borrowers in the early 2000s, many lenders offered enticing ARM mortgages with very low interest rates. These low rates allowed borrowers, even those with lower incomes, to borrow substantial sums to purchase a home. When the interest rates on these mortgages reset in the latter half of the decade, many homeowners were no longer able to afford their payments. Compounding the issue was the fact that home values had fallen so dramatically due to an oversupply in housing inventory from foreclosure. The equity for many homeowners had been wiped out further exacerbating the problem. Stretching to make higher monthly payments on a house in which they held little to no equity proved to be an unwise decision. As a result, many homeowners stopped making monthly mortgage payments. In response, banks and lenders initiated foreclosure proceedings on six million borrowers over this period. The costs of foreclosure for homeowners are substantial – including the loss of one's home, the loss of any equity already put into the home, a tarnished financial record and credit profile, and the need to find new housing.

Prevalence of Foreclosure

In a typical year, a relatively small percentage of homes are foreclosed on. In 2022, the foreclosure rate was just over 0.2% (Statista, 2023). In the immediate aftermath of the subprime mortgage crisis in 2010, the percent of households receiving a foreclosure filing reached an all-time national high of 2.2% – about ten times the level of 2022 (RealtyTrac, 2015) (see Figure 6.1). The years of 2008, 2009, and 2010 each had unprecedented levels of foreclosures as home values fell and mortgage payments increased. At the peak of the crisis, the state with the highest rate of foreclosures was Nevada at nearly 10%. Arizona and Florida also experienced extremely high foreclosure rates of roughly 6% (RealtyTrac, 2010).

Types of Foreclosure Proceedings

The foreclosure process varies by state and generally occurs within three to six months after the first missed mortgage payment. Depending on state law, there are three different types of processes that dictate foreclosure proceedings. A judicial foreclosure provides the borrower 30 days to respond to the demand for payment after the lender initiates the process through the judicial system. If the borrower cannot deliver, the sheriff's office or local court sells the property through an auction to the public. A power of sale foreclosure, which is also called a statutory foreclosure, is allowed if the mortgage stipulates a power of sale clause. The lending institution in this case holds the public auction if the borrower cannot pay. Finally, a strict foreclosure – the least common foreclosure process – involves the lender filing a lawsuit against the borrower; if the borrower cannot repay in the time period stipulated, the property is assumed by the lender.

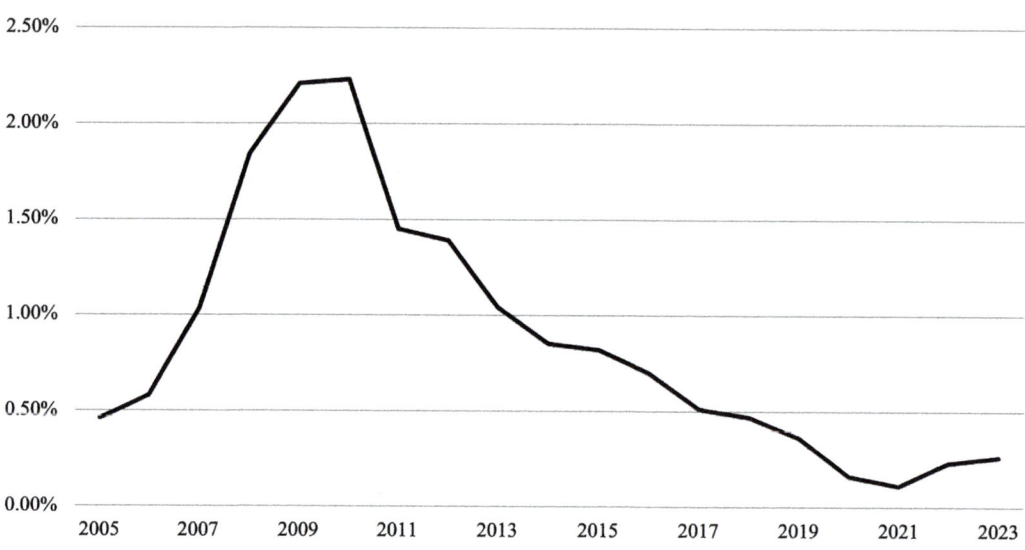

Figure 6.1 Foreclosure Rate (Percent of Housing Units) from 2005–2023
(Source: Attom Data, 2023)

Consequences of Foreclosure

There are numerous adverse consequences of foreclosure. These effects can be felt by both the individual household and the neighborhood in which foreclosures take place. One of the most obvious and intuitive consequences is economic. After a foreclosure, households lose their home and the equity they had invested and built in that home. A foreclosure hinders one's ability to find new housing and damages one's credit scores. These issues can make buying a new home almost impossible, but may also serve as a barrier to finding rental housing (Martin, 2010). Research also shows that households that experience foreclosure face other economic hardships such as food insecurity (Mykyta, 2015). In addition to foregoing basic health care due to constrained finances, individuals who experience foreclosure face a range of mental health consequences given the stress associated with foreclosure. Higher stress and anxiety are prevalent as are higher levels of depression (Isaacs, 2012).

The consequences are not limited to households that experience foreclosure. The longstanding consequences of the subprime mortgage crisis are still felt today. Foreclosures can lead to vacant properties if there are no immediate buyers for such properties. The presence of vacant properties is associated with an increase in overall crime, and violent crime in particular (Arnio et al., 2012; Boessen & Chamberlain, 2017; Immergluck & Smith, 2006b). Foreclosures can also negatively impact nearby property values (Immergluck & Smith, 2006a). For neighborhoods that were hit hard by foreclosure, the years that followed were difficult as high levels of vacancies produced higher crime and lower home values for residents still living in those neighborhoods. These negative effects explain the attention provided by the federal government to avoid a repeat of the subprime mortgage crisis of the late 2000s.

The federal government took steps to mitigate the impacts of the subprime mortgage crisis, including providing mechanisms to allow homeowners to refinance or to modify the terms of their mortgage. But given that these steps were not implemented until after the worst of the crisis, many argued the steps were "too little and too late" (Immergluck, 2013). Rates of foreclosure declined for the remainder of the 2010s as the pressures and dislocations of the housing boom and bust receded. But, in early 2020, the onset of the COVID-19 pandemic created another risk for homeowners. The pandemic caused the most abrupt cessation of economic activity in U.S. history. Unemployment skyrocketed, the stock market crashed, and consumer spending fell dramatically. A concern during this period was homeowners who – due to a loss of income – could no longer make their mortgage payments. To prevent a second financial collapse in just over a decade, the federal government implemented the CARES Act in March 2020. It provided an option for forbearance, in which the lender temporary postpones payments, and prevents a foreclosure by lenders over the term of the forbearance agreement. Over three million mortgages entered forbearance pursuant to the CARES Act (GAO, 2021). The American Rescue Plan, passed in 2021, extended forbearance for all mortgages backed by Fannie Mae and Freddie Mac (GAO, 2021). The foreclosure crisis of the late 2000s cast a long shadow and continued to influence policy over a decade later.

Eviction

In the aftermath of the subprime mortgage crisis of the late 2000s, 3.8 million additional households entered the rental market (Reid et al., 2018). Having lost their homes (and the equity in those homes), purchasing another home was not an option at that time. Further complicating matters, the financial system responded to the lax lending standards of the 2000s by tightening lending requirements to such a degree that it became far more difficult to get a mortgage to purchase a home. As a result, many people who might have otherwise purchased their own home instead remained in or entered the rental market. One anecdote that exemplifies the struggles from that time was the fact that the former Chairman of the Federal Reserve Bank, Ben Bernanke, who helped to oversee the government's response to the subprime mortgage crisis, sought to refinance his mortgage in 2014, but was denied by his bank. In those times of tight credit, the former Princeton economics professor and Chairman of the Federal Reserve did not have sufficient assets and income to encourage his bank to refinance his existing mortgage. The consequences of this denial were likely trivial for Mr. Bernanke, but his experience highlights the extent to which financial conditions had changed in just a few short years.

In the aftermath of the subprime mortgage crisis, there was substantial demand for rental housing in the U.S. The homeownership rate, which had peaked at 69.2% in late 2004, fell to 62.9% by 2015. More people were renting their housing in 2014 than at any time since 1994. As more people entered the rental market, there was limited rental housing available to accommodate the demand. Tight credit conditions limited the financing needed to construct multi-family housing (a primary source of housing for renter households). Constrained building of rental housing exacerbated these tight rental housing conditions. This produced declining rental market vacancy rates and higher rents in the decade following the subprime mortgage crisis. It was a perfect storm of hardship for renters and a windfall for the landlords that were fortunate enough to own rental housing during this period. The decade of the 2010s was marked by substantial increases in rents throughout the nation and increased housing precarity among a large percentage of U.S. renters. In 2015 the average national rent was $959, but by 2021 the average national rent had increased to $1,191 (American Community Survey 1-Year Estimates).

With very low vacancy rates, there was little preventing landlords from raising rents. This was certainly true in growing coastal cities that are known for their high housing costs, such as Boston, San Francisco, and Seattle, but many other communities faced similar pressures in their rental markets. As described elsewhere in this book, a lack of policy support forced many low-income households to find and maintain rental housing in a very unhospitable housing market environment. Given these dynamics, it should come as no surprise that many people were, and remain, unable to afford the ever-increasing rents being charged. When households can no longer afford their rents and stop paying, landlords frequently pursue a legal remedy by filing an eviction notice. An eviction is a formal legal process that allows a landlord to remove a tenant from their unit if they have failed to abide by the terms of the rental contract. Not surprisingly, evictions elicit strong and emotional responses from market actors. Tenants argue that evictions are predatory and place households in an impossible position of housing precarity. Landlords argue that eviction is necessary as a means of enforcing the terms of the rental contract.

Prevalence of Eviction

Evicted by Matthew Desmond, published in 2016, brought to light the lives of precariously housed renters. Desmond's narrative not only won the Pulitzer Prize, but also put a human face on the relatively arcane legal process of eviction in the United States. Through Desmond's compelling writing, readers were able to put themselves into the shoes of low-income renters in Milwaukee, Wisconsin who struggled to pay their rent every day. The book skillfully described the downward spiral that ensues after a renter is evicted from their apartment. Finding new housing is a challenge, and even if they find a new job or garner sufficient resources to pay for a new rental unit, subsequent landlords view the record of a prior eviction on one's record as a *scarlet letter* and a disincentive to enter a lease with that person or household.

Based on research from Desmond and the Eviction Lab at Princeton University, we now have a much better understanding of the prevalence and regional distribution of evictions. Between 2000 and 2018, landlords filed 3.6 million eviction cases per annum or a rate of about 7% of all renter households (Gromis et al., 2022). The eviction crisis is not an isolated issue, but rather a common occurrence for many renter households. The eviction filing rate steadily increased from 2000 to 2008 when the housing bubble burst. Since then, the rate has fallen (see Figure 6.2). Nationwide, in 2018 7.8% of renter households received an eviction filing within

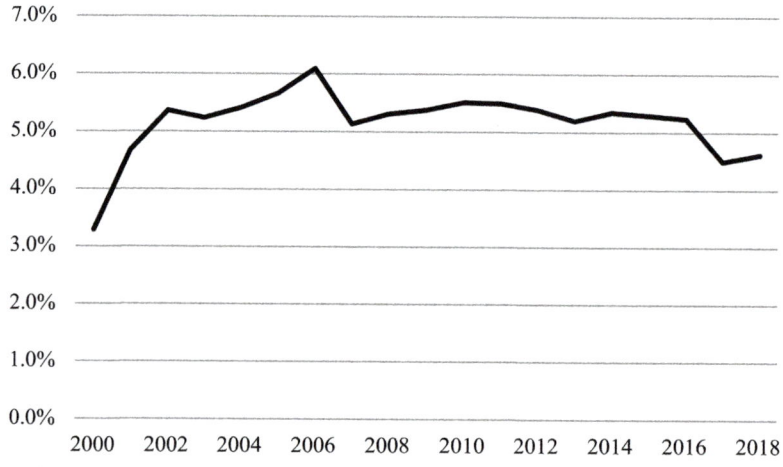

Figure 6.2 National Eviction Filing Rates as a Percent of All Occupied Rental Units per Year from 2000–2018

(Data: Eviction Lab, 2022)

the past year, down from 8.4% in 2015 (Gromis et al., 2022). However, this may be a misleading statistic. The absolute number of eviction filings has continued to grow over this time, but as a percentage of renter households – which we know grew substantially in the aftermath of the subprime mortgage crisis – the overall rate has fallen (Gromis et al., 2022). A preliminary analysis by the Eviction Lab suggests that 2019–2021 eviction rates decreased dramatically due to the federal government's response to the COVID-19 pandemic. However, it remains to be seen how eviction rates will fluctuate once those policies are removed.

Types of Eviction

There are three types of eviction. Just cause, the most common type of eviction, occurs when the tenant has violated the lease. This type of eviction happens if the tenant stops paying rent or violates another term of the lease, such as intentionally damaging the property or involvement in illegal activity in the home. When just cause eviction occurs, landlords must follow formal legal proceedings as dictated by the state, which usually take several weeks to resolve. Jurisdictions that implement just cause evictions require that landlords document the reason for the eviction and adhere to the proper legal process to evict the tenant. Jurisdictions in which most or all rental units are subject to just cause eviction protections include Seattle, Washington DC, and the entire states of California and New Jersey. Research indicates that cities that have adopted just cause evictions have lower eviction filing rates than those that did not (Cuellar, 2019).

Eviction without cause occurs when the landlord evicts a tenant without legal justification. This cannot happen during the term of the lease and occurs if the tenant is still occupying the unit after the lease has expired. Constructive eviction, also referred to as informal eviction, which is the least visible, occurs when the landlord tries to evict a tenant non-legally. For instance, the landlord may defer maintenance and repairs, turn off the electricity, or harass the tenant until they move out on their own. With recent attention on evictions and research that has followed, many jurisdictions have focused on eliminating evictions without cause and constructive evictions.

In addition to just cause eviction policies, there are a range of strategies to protect tenants. To build stronger landlord-tenant relations, landlords may be required to give tenants early notice of eviction with a list of remedies or inform a local government agency so the agency can connect the tenant to resources. Many cities and states are adopting programs to support landlords and provide them incentives to work with tenants. Some local governments have implemented rental registries for oversight of illegal practices such as substantial rent hikes or eviction without cause. Eviction Innovation, a project of Stanford University's Legal Design Lab, hosts a database of these initiatives along with many others that have been implemented to address the eviction crisis.

Consequences of Eviction

Preventing eviction has significant implications for the well-being of the individual and provides broader societal benefits. Evictions are associated with poor physical and mental health for individuals that are forced from their home (Hatch & Yun, 2021). The stress and anxiety generated from housing insecurity, or even the threat of it, can lead to depression and a host of other mental health issues (Acharya et al., 2022; Hatch & Yun, 2021). When an individual is evicted, they are often exposed to temporary, unsafe or over-crowded living situations that impact mental and physical health (Vasquez-Vera et al., 2017). Employment is often interrupted during the eviction process which reduces earnings and destroys credit (Collinson et al., 2023). Reduced earnings and damaged credit further complicate the search for a new unit since the individual may not have funds for a security deposit and landlords are reluctant to rent to individuals with poor credit history and previous evictions. Evictions may also impact democratic participation, depressing voter turnout via several potential factors such as erosion of trust in government among those affected, social isolation, and prioritization of immediate survival (Slee & Desmond, 2021). Most consequentially, eviction is one of the primary factors contributing to homelessness (Collinson et al., 2023; Desmond, 2022; Treglia et al., 2023).

Homelessness

Homelessness represents one of the most obvious and troubling consequences of a lack of stable and affordable housing. In January 2023, more than 650,000 people were estimated to have experienced homelessness on a given night in the United States. But, for those astute observers of homelessness counting and analysis, these estimates are broadly understood to be gross undercounts of the true scale of the crisis. According to the U.S. Department of Housing and Urban Development (HUD), the official definition of homelessness covers an individual or family who lacks a fixed, regular, and adequate nighttime residence, meaning:

(i) Has a primary nighttime residence that is a public or private place not meant for human habitation;
(ii) Is living in a publicly or privately operated shelter designated to provide temporary living arrangements (including congregate shelters, transitional housing, and hotels and motels paid for by charitable organizations or by federal, state, and local government programs); or
(iii) Is exiting an institution where (s)he has resided for 90 days or less and who resided in an emergency shelter or place not meant for human habitation immediately before entering that institution.

It is important to note that there are many precariously-housed people who do not meet this federal definition of homelessness. For example, if you were to lose your housing and live with a friend or relative for a limited period of time (also known as couch surfing or doubling up), you would not be considered homeless per the federal definition, despite lacking permanent and independent housing. Other countries include doubled up households in their national homeless count, but the more limited definition of homelessness in the United States undercounts the number of people who are without stable housing. In addition, the challenges associated with counting the homeless population also explain why the federal point-in-time count of homelessness is probably a significant undercount (Hopper et al., 2008). Each jurisdiction in the United States conducts a point-in-time count at least every other year. During this census, volunteers and local staff members canvas the community to count the number of people residing in homelessness shelters as well as those who are living in unsheltered locations (such as tents, in vehicles, or in a park). The challenge of this exercise helps to explain why published estimates fail to capture the full picture of homelessness in the U.S. The point-in-time count also only captures the people who are currently experiencing homelessness at that specific moment in time. Because people cycle in and out of homelessness, a far greater percentage of the population experiences homelessness over an extended period. For example, Fusaro et al. (2018),

Link et al. (1994), and Tsai (2018) suggest a lifetime prevalence of homelessness between 4 to 7% of the U.S. population.

Among housing-related issues, homelessness tends to be one of the most controversial and polarizing. While people care about homelessness and would like to see it resolved, that is often where alignment ends. Strong views exist on all sides about the root causes of the crisis and how best to resolve it. High rates of homelessness (especially unsheltered homelessness on the West Coast of the U.S.) have a real and profound impact on the people who experience homelessness and the communities in which they reside. There is tremendous pressure on elected leaders to solve the problem, but the resources and political will needed to address the homelessness crisis are frequently lacking. This stalemate creates an environment in which the problem continues to grow while the general public's confidence in society's ability to solve the problem wanes. This troubling quandary has been exacerbated in many cities following the COVID-19 pandemic.

One cannot study and understand homelessness in the U.S. without appreciating the significant racial disparities evident in the population of people experiencing homelessness. People of color are three to four times *over-represented* in the population of people experiencing homelessness compared to their share of the general population (Olivet et al., 2021). This is explained by systemic racism and discrimination in a range of systems including housing, criminal justice, education, the labor market, and healthcare. Hundreds of years of disparities have produced an unequal society in which people of color are substantially over-represented in the homeless population. The systematic changes needed to address the crisis of homelessness in the U.S. must also address racial inequities not just in the housing and homelessness systems, but across society more broadly.

The overall level of homelessness in the U.S. should give us pause. In a nation with substantial wealth and resources, this level of homelessness should be considered unacceptable. Unfortunately, overall rates of homelessness have been relatively stable since the federal government mandated an annual census of the homeless population in 2007. What has changed over that period is the distribution of homelessness throughout the country. While overall levels have fallen slightly, homelessness has become more pronounced and prevalent in a relatively small number of metropolitan areas. There are elevated levels of homelessness in coastal cities such as Boston, New York, Washington, DC, Seattle, Portland, San Francisco, and Los Angeles. Each of these communities have per capita rates of homelessness nearly five times that of cities like Indianapolis and Chicago (Colburn & Aldern, 2022). Why is homelessness so much more prevalent in our coastal cities?

Causes of Homelessness

One of the fundamental challenges in understanding the causes of homelessness is that the answer to this question depends upon the unit of analysis. In research terms, unit of analysis refers to the specific subject, object, or entity of the research such as an individual, household, group of people, neighborhood, or an event, to name a few examples. In the case of homelessness, we could ask, *what causes homelessness at the individual level?* In other words, what are the risk or causal factors for homelessness for an individual? Poverty (Shelton et al., 2009; Tsai, 2018), household composition (Early, 2005), limited social networks (Corinth & Rossi-de Vries, 2018; LaGory et al., 1991), sexual orientation (Castellanos, 2016), substance use and behavioral health challenges (Greenberg & Rosenheck, 2010; Shelton et al., 2009; Tsai, 2018; Early, 2005), having a criminal record (Shinn & Khadduri, 2020), and race (Early, 2004; Olivet et al., 2021) all contribute to an increased risk of homelessness. An individual with any of these conditions, experiences, or identities will be more likely to experience homelessness. Many people center these individual factors when they consider the causes of homelessness.

If we ask the question in a different way, we could instead inquire, *what causes homelessness at the community level?* There are a range of structural explanations for homelessness that go beyond the individual risk factors outlined in the prior paragraph. Among these explanations are economic inequality (Byrne et al., 2021) and systemic racism (Olivet et al., 2021). We can further highlight how structure explains homelessness by asking, *what explains the huge variation in rates of homelessness in cities across the United States?* The high rates of homelessness in New York, Los Angeles, and San Francisco cannot be attributed to more poor people, more addicts, or more people experiencing mental illness in those communities. The reason that coastal cities have high rates of homelessness is due to housing market conditions, and specifically the high cost of housing and lack of affordable housing options (Colburn & Aldern, 2022). Extensive research shows a causal relationship between high housing costs and high rates of homelessness in a community (Hanratty, 2017; Kang, 2019; Byrne et al., 2013; Fargo et al., 2013; Quigley & Raphael, 2001). Places with high rents and low rental vacancies tend to have far higher rates of homelessness than communities with more accessible (and affordable) housing. The key to this argument is not that drugs and mental illness don't matter – they certainly do. The point is that there are people with a wide range of vulnerabilities in every community in the nation. The *difference* is the context in which people live. A person living below the federal poverty line may be able to secure housing in St. Louis where median rents are about $700 per month, whereas a similarly poor person in Seattle has little chance of procuring housing. A rather confounding statistic is that the communities with the highest levels of poverty tend to have relatively low levels of homelessness. This is a head scratcher for many because, clearly, poverty causes homelessness at the individual level. The data demonstrate, however, that homelessness tends to thrive in affluent communities like San Francisco and Seattle, and not in more impoverished cities such as Detroit and St. Louis. This somewhat complicated narrative about the causal drivers of homelessness helps to explain why people frequently misdiagnose the reasons that certain locations have more homelessness than others.

Types of Homelessness

The federal government publishes annual reports that break the homeless population down into constituent parts. For example, an obvious distinction exists between the sheltered and unsheltered population. According to the 2022 Annual Homelessness Assessment Report (AHAR) to Congress, roughly 60% of the people experiencing homelessness were residing in sheltered locations – homeless shelters, transitional housing, or safe havens. The remaining 40% live in unsheltered locations – on the street, in encampments, or in vehicles (de Sousa et al., 2022). But these national figures mask significant variation in the composition of homelessness across different communities. In East Coast cities, such as New York and Boston, the vast majority (over 95%) of people experiencing homelessness are sheltered. These cities have constructed robust shelter systems that provide shelter for most people experiencing homelessness. Cities on the West Coast have not constructed large shelter systems and, as a result, West Coast cities have a roughly even split between sheltered and unsheltered homelessness. This is important because unsheltered homelessness tends to be more visible, while sheltered homelessness is hidden from the public's view. Therefore, people might erroneously think that the problem of homelessness in Los Angeles is far greater than in New York because the crisis is so much more visible in Los

Angeles. While incorrect, this is a widely held belief based on personal experience and anecdote. The challenge for West Coast cities is to determine the appropriate response to the problem of homelessness. Should they construct more shelter capacity to "house" the unsheltered population? This doesn't end homelessness, it simply recharacterizes it (from unsheltered to sheltered) and hides it from view. As there is immense public pressure on elected leaders to deal with the crisis of unsheltered homelessness, this is a tempting path. The consequence of this approach is that it takes time, energy, and resources away from building permanent housing which is the best path to ending homelessness for people experiencing homelessness. This is a central tension in many U.S. cities.

The population of people experiencing homelessness can also be broken down by one's experience of homelessness. Scholars have noted that many people experiencing homelessness have relatively short spells, in other words, their experiences are transitional or episodic. For a smaller percentage of the population, their experiences are more chronic (Kuhn & Culhane, 1998; Nooe & Patterson, 2010). Because people experiencing chronic homelessness tend to have higher needs and consume greater resources, this sub-population receives disproportionate attention from the homeless response system, civic leaders, and the media. But it is important to note that the chronically unsheltered person on the streets who is addicted or mentally ill represents only a small subset of the overall population. Many more people reside in shelters or their cars while they deal with relatively short spells of homelessness before returning to a more stable residential situation. In 2022, 30% of the homeless population was categorized as chronic. Per the federal definition, a chronically homeless individual refers to

> an individual with a disability who has been continuously homeless for one year or more or has experienced at least four episodes of homelessness in the last three years where the combined length of time homeless on those occasions is at least 12 months.
>
> (de Sousa et al., 2022)

Because sheltered homelessness is hidden from view, many people form their opinions about the composition of the homeless population based on who they see – frequently those experiencing chronic, unsheltered homelessness. As the data clearly demonstrate, this is not an accurate depiction. Just under 30% of people experiencing homelessness include families with children. Overall, 23.7% of the homeless population is under the age of 25, highlighting that children and youth are a significant percentage of the overall homeless population. On the other end of the age spectrum is the growing problem of senior homelessness. Recent studies highlight that people over 55 years old are the fastest growing segment of the homeless population and projections suggest a significant increase in elderly homelessness by 2030 (Culhane et al., 2019; Kushel & Moore, 2023). This growth is troubling because of the significant health challenges associated with aging and homelessness. Combining the two is recipe for significant health challenges for older people experiencing homelessness.

The most significant, and alarming, demographic analysis related to homelessness concerns race and ethnicity. People of color are three to four times overrepresented in the homeless population (Olivet et al., 2021; de Sousa et al., 2022). Historic disparities across multiple systems, including housing, criminal justice, education, and healthcare have produced the gross racial and ethnic disparities among people experiencing homelessness. As a result, governments and non-profits focused on addressing homelessness have centered racial justice in their work to honor and acknowledge how systemic racism, present since the founding of the U.S., has contributed to the crisis of homelessness.

Consequences of Homelessness

The consequences of homelessness are significant. Even short exposure to homelessness can have profound consequences for people, especially children (Buckner, 2008; Cutuli et al., 2013). Prolonged experiences with homelessness are associated with far higher rates of mortality (Brown et al., 2022; Roncareti et al., 2018), a range of poor health outcomes (Kushel et al., 2002), high rates of physical and sexual assault (Kushel et al., 2003), and a range of mental health issues, including depression and addiction (Johnson & Chamberlin, 2008). These costs exact a huge human and financial toll. A more humane, or fiscally wise, approach to the problem would be to use the housing interventions that have been empirically demonstrated to end homelessness, such as Housing First (Aubry et al., 2015; Padgett et al., 2016), permanent supportive housing (Corinth, 2017; Rog et al., 2014), and rental subsidies (Gubits et al., 2018). While these programs can be expensive, they are frequently cheaper than the costs associated with homelessness. Homelessness exacts significant costs on a wide variety of community systems, including emergency room, public health, public safety, streets and sanitation, libraries, and criminal justice. Estimates suggest that the annual cost of homelessness for a community is between $30,000 and $100,000 per person (Ly & Latimer, 2015), while the cost savings associated with supportive housing interventions reduced annual costs by between $35,000 and $50,000 per person (Parsell et al., 2017). None of these responses are inexpensive, but one solution houses people, while the other lets homelessness continue to grow and flourish with significant human, social, and financial costs.

Homelessness Interventions

The best solution to homelessness is housing. There is abundant evidence to demonstrate that access to housing ends homelessness. However, like most issues in the social world, the details are more complicated. For some people, receiving a housing voucher, which allows them to access housing in the private market, is all that is needed to end their experience of homelessness. For example, in the Family Options Study, families that received a long-term housing subsidy – in the form of a voucher – had more favorable outcomes than the other interventions that were tested (Gubits et al., 2018). Other people may need housing paired with supportive services. This could include a range of interventions such as transitional housing or permanent supportive housing. In this case, people can receive housing (either temporarily or permanently) while receiving the supportive services they need to deal with other conditions, such as substance use disorders or mental illness.

There is a robust literature demonstrating that the Housing First intervention (which began in the early 1990s) has had great success in ending homelessness for people who experience chronic homelessness (Aubry et al., 2015; Padgett et al., 2016). The idea behind Housing First is to provide what a homeless individual needs when they need it. If the person needs a stable home before tackling other co-occurring conditions, such as addiction, a home is provided. Once the person is stably housed, then they can begin to deal with other challenges they face. Housing First has been shown to work for different types of people in different geographic contexts. Despite robust opposition to the Housing First model, the data demonstrate that it is the single best way to resolve chronic homelessness. Opponents tend to make arguments such as, "If Housing First works so well, why do we still have such high levels of homelessness?" The proper response to such critiques is that society's failure is not due to the ineffectiveness of the intervention, rather the inability and unwillingness to scale these interventions. No matter how effective an intervention is, failure to scale that intervention will not make the lasting impact that is desired.

More broadly, there is a growing understanding that simple access to safe and affordable housing is a critical ingredient to both ending and preventing homelessness. Access to affordable housing can be provided through a range of different mechanisms. First, it might be naturally occurring. For example, market rate rents in cities like Detroit, Cleveland, and St. Louis remain well below $1,000 a month. As a result, homelessness is less prevalent in these communities given the relatively accessible stock of housing. For communities that do not enjoy cheap housing, there are a range of housing supports (discussed at length throughout this book) for low-income households. The limited nature of these housing programs and supports helps to explain the significant level of housing precarity experienced by millions of households across the United States. Homelessness is one significant manifestation of these unmet housing needs. Therefore, while the focus of this book is affordable housing, one could also think about the approaches described in these chapters as ways to enhance housing security and reduce levels of homelessness.

Conclusion

One of the primary reasons for the intense focus on the topic of affordable housing in the United States is the clear evidence about the consequences of unstable and unaffordable living arrangements. Housing precarity is associated with a range of negative outcomes and one of the easiest ways to limit or prevent such precarity is to ensure that households have access to housing that is safe and affordable. Foreclosure, eviction, and homelessness are all consequences of housing precarity that have serious consequences individually, but we also know that these phenomena have a causal relationship. Research clearly demonstrates that eviction can cause homelessness in many circumstances. Therefore, efforts to reduce eviction and foreclosure and improve housing stability will provide a range of household benefits, including a reduction in rates of homelessness. If the nation continues to have an inadequate response to the affordable housing crisis, its negative byproducts – including foreclosure, eviction, and homelessness – will persist.

References

Aborode, A. T. (2022). Threats of evictions in the USA: A public health concern. *Annals of Medicine and Surgery*, *82*, 104681.

Acharya, B., Bhatta, D., & Dhakal, C. (2022). The risk of eviction and the mental health outcomes among the US adults. *Preventive Medicine Reports*, 101981.

Arnio, A. N., Baumer, E. P., & Wolff, K. T. (2012). The contemporary foreclosure crisis and US crime rates. *Social Science Research*, *41*, 1598–1614.

Aubry, T., Nelson, G., & Tsemberis, S. (2015). Housing First for people with severe mental illness who are homeless: A review of the research and findings from the At Home-Chez Soi demonstration project. *The Canadian Journal of Psychiatry*, *60*(11), 467–474.

Boessen, A., & Chamberlain, A. W. (2017). Neighborhood crime, the housing crisis, and geographic space: Disentangling the consequences of foreclosure and vacancy. *Journal of Urban Affairs*, *39*(8), 1122–1137.

Brown, R. T., Evans, J. L., Valle, K., Guzman, D., Chen, Y., & Kushel, M. (2022). Mortality among homeless older adults: Findings from the HOPE HOME prospective cohort study. *JAMA Internal Medicine*, *182*(10), 1052–1060.

Buckner, J. (2008). Understanding the impact of homelessness on children: Challenges and future research directions. *American Behavioral Scientist*, *51*(6), 721–736.

Byrne, T., Harwood, B. F., & Orlando, A. W. (2021). A rising tide drowns unstable boats: How inequality creates homelessness. *Annals of the American Academy of Political and Social Science*, *693*(1), 28–45.

Byrne, T., Munley, E. A., Fargo, J. D., Montgomery, A. E., & Culhane, D. P. (2013). New perspectives on community-level determinants of homelessness. *Journal of Urban Affairs, 35*(5), 607–625.

Castellanos, H. D. (2016). The role of institutional placement, family conflict, and homosexuality in homelessness pathways among Latino LGBT youth in New York City. *The Journal of Homosexuality, 63*(5), 601–632.

Colburn, G., & Aldern, C. P. (2022). *Homelessness is a housing problem: How structural factors explain U.S. patterns.* Oakland, CA: University of California Press.

Collinson, R., Humphries, J. E., Mader, N. S., Reed, D. K., Tannenbaum, D. I., & van Dijk, W. (2023). *Eviction and poverty in American cities.* Working Paper. National Bureau of Economic Research. Retrieved from www.nber.org/papers/w30382

Corinth, K. (2017). The impact of permanent supportive housing on homeless populations. *Journal of Housing Economics, 35*, 69–84.

Corinth, K., & Rossi-de Vries, C. (2018). Social ties and the incidence of homelessness. *Housing Policy Debate, 28*(4), 592–608.

Cuellar, J. (2019). Effect of "just cause" eviction ordinances on eviction in four California cities. *Journal of Public & International Affairs.* Retrieved from https://jpia.princeton.edu/news/effect-just-cause-eviction-ordinances-eviction-four-california-cities

Culhane, D., Doran, K., Schretzman, M., Johns, E., Treglia, D., Byrne, T., Metraux, S., & Kuhn, R. (2019). The emerging crisis of aged homelessness in the US: Could cost avoidance in health care fund housing solutions? *International Journal of Population Data Science, 4*(3), 024.

Cutuli, J. J., Desjardens, C. D., Herbers, J. E., Long, J. D., Heistad, D., Chan, C., Hinz, E., and Masten, A. S. (2013). Academic achievement trajectories of homeless and highly mobile students: Resilience in the context of chronic and acute risk. *Child Development, 84*(3), 841–857.

de Sousa, T., Andrichik, A., Cuellar, M., Marson, J., Prestera, E., & Rush, K. (2022). *The 2022 annual homelessness assessment report (AHAR) to Congress. Part 1: Point-in-time estimates of homelessness.* The U.S. Department of Housing and Urban Development: Office of Community Planning and Development.

Desmond, M. (2022). Unaffordable America: Poverty, housing, and eviction. In E. Mueller & R. Tighe (Eds.), *The affordable housing reader,* 2nd edition (pp. 389–395). Routledge.

Early, D. W. (2004). The determinants for homelessness and the targeting of housing assistance. *Journal of Urban Economics, 55*(1), 195–214.

Early, D. W. (2005). An empirical investigation of the determinants of street homelessness. *Journal of Housing Economics, 14*(1), 27–47.

Eviction Lab. (2022). *Eviction Lab Data Downloads.* Retrieved from https://data-downloads.evictionlab.org/#estimating-eviction-prevalance-across-us/

Fargo, J. D., Munley, E. A., Byrne, T. H., Montgomery, A. E., & Culhane, D. P. (2013). Community-level characteristics associated with variation in rates of homelessness among families and single adults. *American Journal of Public Health, 103*(Suppl 2), 340–347.

Fusaro, V. A., Levy, H. G., & Shaefer, H. L. (2018). Racial and ethnic disparities in the lifetime prevalence of homelessness in the United States. *Demography, 55*, 2119–2128.

Greenberg, G. A., & Rosenheck, R. A. (2010). Mental health correlates of past homelessness in the National Comorbidity Study Replication. *Journal of Health Care for the Poor and Underserved, 21*(4), 1234–1249.

Gromis, A., Fellows, I., Hendrickson, J. R., Edmonds, L., Leung, L., Porton, A., & Desmond, M. (2022). Estimating eviction prevalence across the United States. *Proceedings of the National Academy of Sciences of the United States of America, 119*(21), 1–8.

Gubits, D., Shinn, M., Wood, M., Brown, S. R., Dastrup, S. R., & Bell, S. H. (2018). What interventions work best for families who experience homelessness? Impact estimates from the Family Options Study. *Journal of Policy Analysis and Management, 37*(4), 835–866.

Hanratty, M. (2017). Do local economic conditions affect homelessness? Impact of area housing market factors, unemployment, and poverty on community homeless rates. *Housing Policy Debate, 27*(4), 640–655.

Hatch, M. E., & Yun, J. (2021). Losing your home is bad for your health: Short- and medium-term health effects of eviction on young adults. *Housing Policy Debate, 31*(3-5), 469–489.

Hopper, K., Shinn, M., Laska, E., Meisner, M., & Wanderling, J. (2008). Estimating numbers of unsheltered homeless people through plant-capture and postcount survey methods. *Innovations in Design and Analysis, 98*(8), 1438–1442.

Immergluck, D. (2013). Too little, too late, and too timid: The Federal response to the foreclosure crisis at the five-year mark. *Housing Policy Debate, 23*(1), 199–232.

Immergluck, D., & Smith, G. (2006a). The external costs of foreclosure: The impact of single-family mortgage foreclosures on property values. *Housing Policy Debate, 17*(1), 57–79.

Immergluck, D., & Smith, G. (2006b). The impact of single-family mortgage foreclosures on neighborhood crime. *Housing Studies, 21*(6), 851–866.

Isaacs, J. B. (2012). *The Ongoing Impact of Foreclosures on Children.* Washington, DC: Brookings.

Johnson, G., & Chamberlain, C. (2008). Homelessness and substance abuse: Which comes first? *Australian Social Work, 61*(4), 342–356.

Kang, S. (2019). Beyond households: Regional determinants of housing instability among low-income renters in the United States. *Housing Studies, 36*(1), 80–109.

Kuhn, R., & Culhane, D. P. (1998). Applying cluster analysis to test a typology of homelessness by pattern of shelter utilization: Results from the analysis of administrative data. *American Journal of Community Psychology, 26*(2), 207–232.

Kushel, M. B., Perry, S., Bangberg, D., Clark, R., & Moss, A. R. (2002). Emergency department use among the homeless and marginally housed: Results from a community-based study. *American Journal of Public Health, 92*(5), 778–784.

Kushel, M. B., Evans, J. L., Perry, S., Robertson, M. J., & Moss, A. R. (2003). No door to lock: Victimization among homeless and marginally housed persons. *Archives of Internal Medicine, 163*(20), 2492–2499.

Kushel, M., & Moore, T. (2023). *Toward a new understanding: The California statewide study of people experiencing homelessness.* Benioff Homelessness and Housing Initiative, University of California San Francisco.

LaGory, M., Ritchey, F., & Fitzpatrick, K. (1991). Homelessness and affiliation. *The Sociological Quarterly, 32*(2), 201–218.

Link, B. G., Susser, E., Stueve, A., Phelan, J., Moore, R. E., & Struening, E. (1994). Lifetime and five-year and lifetime prevalence of homelessness in the United States. *American Journal of Public Health, 84*(12), 1907–1912.

Ly, A., & Latimer, E. (2015). The impact of federal homelessness funding on homelessness. *Southern Economic Journal, 84*(2), 548–576.

Makridis, C., & Ohlrogge, M. (2021). Moving to opportunity? The geography of the foreclosure crisis and the importance of location. *Journal of Economic Geography, 22*(1), 159–180.

Martin, A. J. (2010). After foreclosure: The displacement crisis and the social and spatial reproduction of inequality. *Institute for the Study of Social Change Working Paper Series 2009–2010.*

Mykyta, L. (2015). Housing crisis and family well-being: Examining the effects of foreclosure on families. *SEHSD Working Paper #2015–07.*

Nooe, M., & Patterson, D. A. (2010). The ecology of homelessness. *Journal of Human Behavior in the Social Environment, 20*(2), 105–152.

Olivet, J., Wilkey, C., Richard, M., Dones, M. Tripp, J., Beit-Arie, M., Yampolskaya, S., & Cannon, R. (2021). Racial inequality and homelessness: Findings from the SPARC Study. *The Annals of the American Academy of Political and Social Sciences, 693*(1), 82–100.

Parsell, C., Peterson, M., & Culhane, D. (2017). Cost offsets of supportive housing: Evidence for social work. *British Journal of Social Work, 47*(5), 685–689.

Quigley, J. M., & Raphael, S. (2001). The economics of homelessness: The evidence from North America. *International Journal of Housing Policy, 1*(3), 323–336.

Padgett, D. Henwood, B. F., & Tsemberis, S. J. (2016). *Housing First: Ending Homelessness, Transforming Systems, and Changing Lives.* Oxford: Oxford University Press.

RealtyTrac. (2010). *Year-End Report.* Irvine, CA: RealtyTrac. Retrieved from www.documentcloud.org/documents/23343754–2010–01–14-realtytrac-year-end-report-shows-record-28-million-us-properties-with-foreclosure-filing

RealtyTrac. (2015). *U.S. foreclosure starts increase to 17-month high in December; Average time to foreclose nationwide decreases for first time since Q1 2011.* Irvine, CA: RealtyTrac.

Reid, C. K., Sanchez-Moyano, R., & Galante, C. J. (2018). *The rise of single-family rentals after the foreclosure crisis: Understanding tenant perspectives.* Terner Center, CA. Retrieved from https://ternercenter.berkeley.edu/wp-content/uploads/pdfs/Single-Family_Renters_Brief.pdf

Rog, D. J., Marshall, T., Dougherty, R. H., George, P., Daniels, A. S., Ghose, S. S., & Delphin-Rittmon, M. E. (2014). Permanent supportive housing: Assessing the evidence. *Psychiatric Services, 65*(3), 287–294.

Roncareti, J. S., Baggett, T. P., O'Connell, J. J., Hwang, S. W., Cook, E. F., Krieger, N., & Sorensen, G. (2018). Mortality among unsheltered homeless adults in Boston, Massachusetts, 2000–2009. *JAMA Internal Medicine*, *178*(9), 1242–1248.

Shelton, K. H., Taylor, P. J., Bonner, A., & van den Bree, M. (2009). Risk factors for homelessness: Evidence from a population-based study. *Psychiatric Services*, *60*(4), 465–472.

Shinn, M., & Khadduri, J. (2020). *In the midst of plenty: How to prevent and end homelessness*. Hoboken, NJ: Wiley.

Slee, G., & Desmond, M. (2021). Eviction and voter turnout: The political consequences of housing instability. *Politics & Society*, *51*(1).

Statista. (2023). *Foreclosure Rate in the United States from 2005 to 2022*. Retrieved from www.statista.com/statistics/798766/foreclosure-rate-usa/

Treglia, D., Byrne, T., & Tamla Rai, V. (2023). Quantifying the impact of evictions and eviction filings on homelessness rates in the United States. *Housing Policy Debate*.

Tsai, J. (2018). Lifetime and 1-year prevalence of homelessness in the US population: Results from the National Epidemiologic Survey on Alcohol and Related Conditions-III. *Journal of Public Health*, *40*(1), 65–74.

U.S. Census Bureau and U.S. Department of Housing and Urban Development. (2023). *Average sales price of houses sold for the United States [ASPUS]*. Retrieved from https://fred.stlouisfed.org/series/ASPUS

U.S. Government Accountability Office (GAO). (2021). *Struggling homeowners get extension, but did COVID-19 housing protections really work?* Washington, DC: U.S. Government Accountability Office.

Vasquez-Vera, H., Palencia, L., Magna, I., Mena, C., Neira, J., & Borrell, C. (2017). The threat of home eviction and its effects on health through the equity lens: A systematic review. *Social Science & Medicine*, *175*, 199–208.

Part III

Providing Access to Affordable Housing

Chapter 7

Sectors and Actors Involved in the Provision of Affordable Housing

Overview of Sectors and Actors

Before delving into the details of the affordable housing system in the United States, we place it in a broader global context. Unlike many other countries in the world, the vast majority of housing in the United States is privately developed and operated. Less than 1% of all units are part of the public housing program and only 2.4% of housing is constructed with the support of tax credits (2022 ACS 1-Year Estimates; HUD, 2022; National Housing Preservation Database, 2022). Finally, about 1.9% of housing is occupied by recipients of federal rental assistance in the form of housing vouchers (2022 ACS 1-Year Estimates; HUD, 2022). Even when accounting for all the other smaller HUD programs, only about 5% of housing is supported by the federal government. As a share of the rental stock, the share that are subsidized are higher. For example, the number of Housing Choice Vouchers is equal to 5.6% of all rental housing, public housing is just below 2% of the rental housing stock, and units supported with LIHTC are about 7.1% of all rental housing.

In the face of a national poverty rate of 12.6% (2022 American Community Survey 1-Year Estimates), the limited federal role in affordable housing is conspicuous. State and local governments and non-profit organizations also support the goal of affordable housing through a range of programs and initiatives. The National Low Income Housing Coalition created a Rental Housing Programs Database which identified 353 active rental housing programs across the country, including 281 state-funded programs and 72 that were established by local governments in the nation's largest cities (Abdelhadi & Aurand, 2023). While the scale of these programs (in dollars or number of units supported) is not available, these programs play a significant role in the provision of affordable housing in the nation. To further supplement housing assistance from government sources, the non-profit sector is estimated to account for about 2.3% of the rental housing units in the country (Keightley, 2022), which represents roughly 30% of the units supported by all federal programs. The non-profit sector is, therefore, a major player in the affordable housing system.

The organizations that deliver affordable housing play a diverse range of roles and functions such as developing and managing affordable housing, providing lending services, regulating housing markets, providing advisory and emergency support and services, and delivering non-related housing programs to build economic mobility for residents. The federal government primarily provides federal support through direct subsidies and financing. State government—predominately housing finance agencies—support the development of affordable housing by providing financing for housing and administering federal funds. State governments also operate a range of rental assistance programs that supplement federal efforts. Local governments also provide supports that address deficiencies in federal and state assistance and they are the key players in establishing the regulatory system that oversees the supply of housing. As noted above, the non-profit sector also plays an important role. Nonprofits are often the vehicle for direct delivery of housing and frequently serve as the owner, developer, and manager of moderate- and low-income housing. Nonprofits are unique in that housing is only one aspect of the support structure they provide. For example, some nonprofits construct and operate supportive housing that combines housing with other essential services for people experiencing homelessness, including treatment for behavioral health disorders, job training, and healthcare.

Finally, the private sector also plays a critical role in the delivery of affordable housing. In some cases the private market delivers affordable housing options that are known as Naturally Occurring Affordable Housing (NOAH). NOAH occurs when market conditions align such that housing on the private market is priced such that lower income households can afford it. In a troubling trend, over the last decade, the amount of NOAH in the U.S. has fallen dramatically (Joint Center for Housing Studies, 2023), but the private sector still dominates the housing market in this country and, therefore, is an important player in the provision of affordable housing. There are other ways that private actors help to create affordable housing. Private financial intuitions create programs to expand access to homeownership, private developers collaborate with government in public-private partnerships to produce affordable housing, and private investment is used to increase the availability of affordable housing. For example, the Home Loan Bank System offers liquidity to the mortgage market through loans to financial institutions that supply residential mortgages. Other private actors include LIHTC intermediaries that facilitate the distribution of tax credits to developers and investors, and Community Development Financial Institutions that offer financial services to low-income households.

Overlap between different actors in society is evident in the delivery of affordable housing. A prime example of this multi-sector collaboration is how the Low-Income Housing Tax Credit (LIHTC) program functions. The federal government facilitates this program by providing tax credits that can be used to help fund the construction of affordable housing. The tax credits are distributed via state housing finance agencies. State housing finance agencies design and manage the allocation of tax credits through a competitive process and awards tax credits to both for-profit and non-profit developers. The state housing finance agency may also issue bonds to help finance the construction of housing using tax credits. Finally, local governments provide additional financial support for the development of LIHTC housing in their communities.

Government Agencies

U.S. Department of Housing and Urban Development

The U.S. Department of Housing and Urban Development, often referred to as HUD, is the designated federal agency for housing and community development assistance. In the early 1960s, there was momentum by Democrats to create a new cabinet-level department focused on urban development and housing initiatives, but it wasn't until 1965 under the Johnson Administration that the Federal Housing Administration, the Public Housing Administration, the Urban Renewal Authority, and Fannie Mae were merged to create HUD. HUD's main mission is to support and facilitate housing and community development. Over the last five decades, HUD has insured over 44 million home mortgages, provided rental housing assistance (including public housing) to over 35 million individuals, invested approximately $150 billion through the Community Development Block Grant program, provided more than $14 billion to address homelessness, helped over 90,000 individuals combat housing discrimination with over $150 million awarded in compensation, and developed over 85,000 housing units for Native Americans (Khadduri, 2015).

While HUD plays a primary role in federal housing policy, there are other departments and agencies that are actively involved. For example, the U.S. Interagency Council on Homelessness helps to coordinate the federal response to homelessness; home loans for veterans are provided by the Department of Veterans Affairs; the LIHTC program is administered by the Department of Treasury; and rural housing is supported via the Department of Agriculture, Rural Housing Services division. As this list highlights, responsibility for housing support in the United States is not centralized in one agency, which adds complexity to the process. In addition, politics affects how HUD is managed and the resources devoted to these programs. Because support for the programs delivered by HUD vary across presidential administrations, advocacy for HUD and the programs it delivers changes over time. These changes complicate efforts around long-term planning of housing support. In fact, HUD budgetary authority dropped meaningfully over the last several decades, most notably during Ronald Reagan's presidency in the early 1980s.

An ongoing challenge for HUD is that many social problems have fallen to them, including urban poverty and residential segregation. While each of these social ills have a significant housing element, the scale of the challenges exceeds the budget and programmatic purview of the agency. HUD, for example, is unable to eliminate discrimination in the housing market – federal or state prosecutors respond to cases of discrimination – nor is HUD able to address the inadequate wages that cause many households to experience high levels of housing cost burden. While HUD may not have primary responsibility for addressing these societal ills, the agency is not without blame. There are well-documented challenges with HUD's management of the public housing program. In addition, HUD's past historical emphasis on urban renewal and slum clearance has drawn understandable critiques. These issues have led to major debates about the scope and role of the federal government in building and owning housing, revitalization efforts, and gentrification. As described elsewhere in this book, an outcome of these debates was the devolution of responsibility for housing support. Current programs push decision-making and oversight for federal programs to local agencies and governments. Given the inadequacy of the federal response to housing need – driven by inadequate funding – many other players have stepped into the breach to attempt to fill this need.

Government-Sponsored Enterprises

Government-sponsored enterprises (GSEs) are federally chartered to provide stability and liquidity to the housing finance market. Federal National Mortgage Association (Fannie Mae) and Federal Home Loan Mortgage Corporation (Freddie Mac) are the two most recognized GSEs. Fannie Mae was created in 1938 as part of the New Deal and Freddie Mac in 1970 as a competitor to Fannie Mae in the secondary mortgage market. Both agencies buy and securitize mortgages from lenders to infuse liquidity in the mortgage market which increases the number of loans lenders can offer. During the 2008 financial collapse, Fannie Mae and Freddie Mac faced severe financial challenges as a result of significant financial losses due to the number of delinquent mortgage loans in their portfolios.

The Housing and Economic Recovery Act of 2008 created the Federal Housing Finance Agency (FHFA) to oversee and regulate Fannie Mae, Freddie Mac, and the Federal Home Loan Bank System, legally referred to as conservatorship. In the last few years there has been significant discussion on reducing the role of the FHFA and privatizing Fannie Mae and Freddie Mac but both GSEs currently remain under government conservatorship. The primary role of the FHFA is to ensure there is a reliable source of liquidity and funding for housing finance and community investment. These basic functions were disrupted during the Global Financial Crisis of 2007 – 2008 and there was a desire by the government to ensure ongoing healthy functioning of the housing finance industry. The FHFA serves to promote equitable access to affordable housing. The agency engages in regulation and enforcement, initiatives and programs, outreach and education, and research to achieve this goal. FHFA also oversees the Federal Home Loan Bank System which provides funding for mortgages to moderate and low-income households and supports community development activities and affordable housing projects in under-resourced communities. It ensures that these agencies comply with affordable housing statutory obligations. Finally, the FHFA has several programs and initiatives that provide loan modification and other assistance to homeowners who are struggling to stay in their homes.

U.S. Interagency Council on Homelessness

The U.S. Interagency Council on Homelessness (USICH) works with other federal agencies and state and local governments across the country to combat homelessness. Through facilitation and partnership, USICH engages government agencies, scholars, non-profit organizations, the private sector, and philanthropic entities to conduct research, establish policy, and launch initiatives that leverage resources and expertise to prevent and end homelessness.

The USICH is part of the executive branch of the federal government, but it is not its own agency and is not represented in the President's Cabinet. Instead, USICH interacts with the Secretaries of other federal agencies, such as HUD, Department of Health and Human Services, and the Veterans Affairs

> to coordinate the federal response to homelessness and to create a national partnership at every level of government and with the private sector to reduce and end homelessness in the nation while maximizing the effectiveness of the federal government in contributing to the end of homelessness.
> (United States Interagency Council on Homelessness, 2024)

In 2024, the council was chaired by the Secretary of Health and Human Services Secretary and the Vice Chair was the Secretary of the Department of Agriculture. Leadership of USICH rotates among members of the President's Cabinet. By establishing the USICH in 1987, the government acknowledged the challenging nature of the problem of homelessness; ending homelessness would require cross-agency collaboration given the complexity of the problem.

Department of Veterans Affairs

In addition to the well-known provision of healthcare, the Department of Veterans Affairs (VA) also plays an important role in the provision of affordable housing to veterans and their families. A primary mechanism used to facilitate access to housing is through the VA Home Loan Program. The mortgages offered through this program require no down payment and no private mortgage insurance. These mortgage loans contain lower interest rates and reduced closing costs to provide active-duty service members, veterans, and surviving spouses access to affordable homeownership. The VA also supports affordable rental housing through programs like Veterans Affairs Supportive Housing (VASH) which provides Housing Choice Vouchers to veterans and supports housing developments that increase the supply of units available to veterans.

For example, the Enhanced-Use Lease (EUL) program allows the VA to enter long-term leasing arrangements on underutilized properties for the purpose of housing low-income veterans. The VA also plays a significant role in providing housing and supportive services for veterans and their families that are experiencing or at risk of homelessness. The VA has been celebrated for its work in ending veteran homelessness. In the ten years ending in 2019, the number of veterans experiencing homelessness in the United States was cut in half.

USDA Rural Housing Services

As a division of the Department of Agriculture, Rural Housing Services (RHS) provides support for affordable housing in rural areas by providing loans with generous terms. RHS also purchases, builds, and renovates single-family homes in rural areas, and supports multi-family housing development and improvement. The division also supports infrastructure in rural communities that is vital to the quality of life for residents by providing loans and grants to improve water quality and infrastructure, and to construct basic community services such as schools, fire stations, and health care facilities.

State Housing Finance Agencies

State housing finance agencies, referred to as HFAs, are a key financial player in the housing delivery system. State housing finance agencies' activities include selling bonds to finance the development of housing, acting as direct lenders to affordable multi-family housing developers, administering federally funded programs such the Low-Income Housing Tax Credit program, the HOME Investment Partnership Program, and Community Development Block Grants, offering foreclosure prevention and counseling programs, advocating for policies and regulation that promote the development and preservation of affordable housing, performing state housing planning activities, and generating and managing dedicated sources of funding for housing activities through housing trust funds.

Created by state enabling statute, HFAs operate under a board of directors that sets policies. The board is appointed by an elected official, typically the state governor. The board appoints an executive director who hires other professional staff. While all states have a housing finance agency, the scope and functions may vary. A variety of factors may influence what an agency does and where it may focus its efforts, including a state's housing priorities and needs, local housing market conditions, population, resources and funding, and relevant state laws. Because of these factors, the same programs may be implemented differently across states. Differences may include program generosity and eligibility. In some cases, state housing finance agencies also assume the role traditionally held by public housing authorities, such as administering the Housing Choice Voucher program. For example, the Alaska Housing Finance Commission administers vouchers for 12 communities throughout the state and the Delaware State Housing Authority serves as both the state housing finance agency and public housing authority.

State and Local Government

Because the federal government has played an increasingly limited role in supporting affordable housing over the last half century, there has been increased pressure on state and local governments to address the unmet need. Only one in four people who are eligible for federal housing support actually receives it, therefore, the other three in four households turn to state and local governments for assistance. Because housing is such a significant cost for households, attempts to meet these needs place significant burdens on local governments. As affordable housing challenges spread throughout the United States, we see state and local governments playing visible and important roles.

State governments engage with the federal government, counties, municipalities, rural areas, non-profits, and the private sector to expand the affordable housing stock. State governments also play a vital role in regulating local housing markets in their states. These laws can both promote the expansion of affordable housing within the state or exacerbate the scarcity of affordable housing. For instance, land use and zoning regulations and the ability for local government to enact inclusionary zoning, tax increment financing, and rent control programs impact the provision of affordable housing. As an example, in Texas, municipalities are only allowed to implement rent control measures if a housing emergency exists and the governor approves the request whereas California specifically passed a law that encourages and allows rent control ordinances at the local level.

In recent years, states have taken action to combat the affordable housing crisis through legislation. Colorado implemented House Bill 1313, which mandates affordable housing development near public transit. This includes inclusionary zoning and deed restrictions to limit rent and sales prices. Washington passed bills like House Bill 1110 to allow duplexes and fourplexes in most communities, and House Bill 1137 to permit two accessory dwelling units per lot with fewer development restrictions. Oregon's Senate Bill 608 imposed statewide rent regulations, capping annual increases at 7% plus inflation and restricting no-cause evictions. These are a few of many examples across the nation of state legislative efforts to support greater affordable housing.

Local governments have always had a major role in housing provision, primarily by establishing the regulatory environment that governs the development of housing. These rules and regulations include planning and zoning, building codes, the issuance of permits, and enforcement of housing standards and regulations. However, these regulations, sometimes intentionally, have created impediments to ensuring an adequate supply of housing, and particularly affordable housing. As described in Chapter 11, single-family zoning – typically in the purview of local governments – has restricted density and prohibited multi-family housing, thus contributing to increasingly unaffordable housing.

In addition to establishing the rules and regulations that govern the construction of housing, local jurisdictions have also established programs and initiatives to create housing that is more accessible and affordable. Most local governments allocate a portion of their budget or have dedicated funding sources to help low-income households access affordable housing. These programs and initiatives range from land acquisition and development activities to provide support during the development process such as fee waivers and expedited permitting for affordable housing. Some local jurisdictions have very specific programs in place to ensure a certain portion of new residential development is set-aside for lower income households like inclusionary zoning programs that are discussed in detail in the chapter on regulatory strategies.

One example of a city-based initiative focused on affordable housing is the Seattle Housing Levy. The City of Seattle first passed a housing levy in 1986. In 2023, Seattle renewed the levy for another seven years which totaled $970 million. Funds raised by the housing levy are generated by a property tax, which costs a Seattle homeowner in a median-valued home an additional $383 a year. The funds are used to provide new affordable homes, keep low-income households in their homes, and to prevent homelessness (City of Seattle, 2024). The Seattle Housing Levy highlights how local jurisdictions are not solely dependent upon the federal and state government for funds to support affordable housing.

Public Housing Authorities

As a local extension of HUD, public housing authorities (PHAs) are the primary provider of affordable housing for the lowest income households in the nation. PHAs work directly with residents in their local community to provide public housing, housing vouchers, non-federally subsidized affordable housing units, and social service resources. For example, New York City Housing Authority (NYCHA) alone provides housing or housing assistance to over 500,000 individuals serving about 1 in every 17 New Yorkers. Furthermore, NYCHA provides social services, job placement support, and community programing to enhance economic opportunity, wellness, and education attainment for their residents (NYCHA, 2023).

PHAs were created as part of the Housing Act of 1937 and serve as the entities through which federal housing support is distributed. PHAs are quasi-governmental, local agencies that have contracts with HUD known as Annual Contributions Contracts (ACCs). Under these contracts, PHAs receive funding in the form of operating and capital grants and use these funds to administer HUD programs based on federal rules and regulations. However, PHAs are authorized and structured under state law, so PHAs respond to community input and have some discretion to set policy consistent with local priorities.

PHAs typically have an executive director who reports to a governing board that oversees the operations of the organization. The boards may be appointed by local government officials or may be elected. Specifically, the governing board approves policies, helps determine the direction of the PHA, and delegates responsibility and authority to the executive director who acts on its behalf. PHAs are required to have a Resident Advisory Board that participates in the planning process and PHA residents are allowed to elect delegates to represent their interests.

PHAs were originally tasked with administering the federal public housing program. Because of the notoriety of public housing in the United States, this function of PHAs has often overshadowed their many other roles and responsibilities. Not only do PHAs directly administer HUD initiatives such as the Housing Choice Voucher program, but they have also taken a direct role as the leading affordable housing developer in their communities. PHAs own and operate a stock of affordable housing that may, or may not, have federal assistance or other government subsidies tied to it. The activities of PHAs are dictated by their plans, an annual plan and five-year plan, that are created with resident input and require HUD approval. The five-year plan outlines the mission, goals, and objectives of the agency. The annual plan contains the PHA's policies and a summary of day-to-day operations.

There are two assessment tools used to evaluate PHAs. The Public Housing Assessment System (PHAS) assesses a PHA's performance in administering the public housing program, and the Section Eight Management Assessment Program (SEMAP) facilitates oversight for the Housing Choice Voucher program. The PHAS uses several metrics to assess the performance of a PHA, including the physical condition of the housing stock, financial health, management performance, and the use of capital funding by the agency. Overall scores of 90% or higher indicate that an agency is a high performer, which entitles a PHA to additional funding and reduced federal oversight. Troubled agencies, those that score less than 60%, are required to develop plans for improvement. Some troubled agencies may be placed in receivership, in which HUD or another external entity assumes control of the agency to improve its performance. To assess performance in administering the Housing Choice Voucher program, SEMAP relies on 14 different performance indicators, including selection of tenant applications, management of wait lists, accuracy of income verification, and proper calculation of housing assistance payments. Like PHAS, SEMAP scores translate to an overall performance rating (high, standard, and troubled). Corrective action plans are required for troubled agencies and those that are underperforming on any of the indicators.

Non-Profit Organizations

Given the scale of government support for affordable housing, there is a significant gap between what the government provides and the demonstrated need. In response, a range of other actors, including non-profit organizations, have been created to help, in part, address this conspicuous need. Non-profits engage in a variety of activities, including funding and development of affordable housing, managing housing projects and providing supportive services to tenants, and advocating for greater investments in affordable housing and a more robust set of tenant protections for renters. Non-profit organizations are frequently at the forefront of creating affordable housing and delivering housing services. While they may seek their own funding, they often depend on government resources to ensure housing for the lowest income households. There are a wide variety of non-profit organizations involved in the provision of affordable housing, including national non-profits, faith-based organizations, community-based development organizations, non-profit housing developers, grassroots agencies, and philanthropic foundations.

These non-profit organizations have a variety of different roles and functions. Some are member-based organizations that serve a convening function to support education and training, advocacy, and policy work. Examples of these organizations include the National Council of State Housing Agencies, National Association of Housing and Redevelopment Officials, and Public Housing Authorities Directors Association. Other non-profit organizations focus on the development and financing of affordable housing, such as Local Initiatives Support Corporation, Enterprise Community Partners, and NeighborWorks America. These groups promote affordable housing development through financial support, technical assistance, training and grants, and capacity building. Organizations like National Fair Housing Alliance, National Low-Income Housing Coalition, National Housing Conference, and National Alliance to End Homelessness collaborate with industry professionals, policymakers, and tenants to advocate for affordable housing solutions through policy advocacy. Urban Institute, the Joint Center for Housing Studies of Harvard University, the Furman Center at New York University, and the Terner Center for Housing Innovation at University of California Berkely are known nationally as leading research centers on the topic of housing – with a particular focus on affordable housing.

Faith-based organizations have played a central role in the provision of affordable housing since the founding of the United States. In the absence of any formal government responsibility for housing, religious groups have provided shelter for the poor and older adults. The religious community played a prominent role in the 19th and early 20th centuries when poor housing conditions became a central concern in urban cities due to a lack of adequate housing conditions for immigrant populations. Today faith-based organizations continue to provide housing and supportive services. The most well-known example is Habitat for Humanity. Habitat's model is based on sweat equity where the household works alongside a group of volunteers to construct a house in which they will live. Habitat also offers affordable mortgages to provide easier access to homeownership. Several other well-known faith-based organizations that provide affordable housing include the Salvation Army, the Nehemiah Housing Development Fund Corporation, St. Vincent de Paul Society, and Presbyterian Housing.

Community-based development organizations (CBDOs) are non-profit organizations that concentrate on a specific area or neighborhood to provide community development initiatives which often include affordable housing development. One of the most common types of CBDOs are community development corporations. There are over 6,200 CBDOs in the United States (Scally et al., 2023). The impact of CBDOs on housing is noteworthy. According to an Urban Institute report, CBDOs have, over time, produced more than four million units of housing. Beyond housing, CBDOs construct community or commercial facilities, provide credit and loans to local organizations, and provide a range of services and advocacy in the neighborhoods they serve. Funding for CBDOs comes from federal sources such as HOME Investment Partnerships Program (HOME) and Community Development Block Grants (CDBG). Many also receive funding from state and local government sources. The board of directors of CBDOs include community members from the area that understand the needs of the local community. CBDOs range in size from very small to large, sophisticated organizations.

There are numerous mission-driven non-profit housing developers in America that are dedicated to increasing the supply of affordable housing for low-income households. These organizations benefit from being able to access government funding sources to develop affordable housing, including Low-Income Housing Tax Credits, housing trust funds, and Community Development Block Grants which are all discussed in later chapters. There are thousands of local non-profit housing developers in America. Some of the more well-known and largest include BRIDGE Housing Corporation, Mercy Housing, and Stewards of Affordable Housing for the Future.

Grassroots agencies gather around community-based initiatives when a group of residents see an unmet need in their community. Many neighborhood-based organizations have been created to address issues around homelessness and unaffordable housing. Tenant associations are one example and are usually formed to advocate for housing rights. As a group, members can collectively have a voice, organize, share, and advocate for a range of issues, including fair housing policies, affordable rents, and creating safe communities. There are many other non-profit housing models such as co-housing and community land trusts which are discussed in future chapters that also involve grassroots organizing.

Finally, philanthropy plays a vital role in addressing affordable housing challenges. First, foundations provide direct financial support for the development, rehabilitation, and maintenance of affordable housing development. Philanthropy also contributes by funding research to identify best practices and disseminate knowledge on how to address the many challenges that prevent affordable housing. Philanthropic organizations often fund innovative pilots and models that, if successful, can be scaled for greater impact. Last, foundations also support advocacy and policy work along with capacity-building and technical assistance for affordable housing.

Private Sector

The private sector plays a critical role in the delivery of affordable housing. As noted earlier, the private sector constructs a significant portion of housing in the United States. Even publicly-supported affordable housing, built with the Low Income Housing Tax Credit (LIHTC) program, is constructed by private developers. Private sector actors involved in the production of affordable housing include developers, contractors, architects, designers, financial institutions, advisory services, legal and accounting services, private equity investors, and property managers. Developers, contractors, architects, attorneys, and designers all contribute during the development process and often collaborate in public-private partnerships in the production of affordable housing. Private financial institutions offer funding during the construction phase, finance projects once they are in productive use, and create programs to expand access to homeownership for occupants. Advisory services are frequently used by developers of affordable housing given the complexity of the regulations and funding streams involved in those projects. Those who structure these deals often require legal and accounting services to ensure the partnerships are properly structured and comply with tax planning and legal requirements. Private investment is often used to increase the availability of affordable housing. Once developed, the property is often, but not always, managed by for-profit property management companies. There are non-profit developers that also manage properties once they are constructed.

There is a healthy debate about the role of the private sector in the provision of affordable housing. Depending on the local context, private developers – alone – can in certain circumstances construct housing that is affordable for people who earn below the area median income. Of course, local conditions including land, labor, and construction costs as well as the local regulatory environment will impact how deeply affordable a private market developer can construct. In most cases, there is little to no ability for a private developer to construct deeply affordable housing. But even in expensive markets like Seattle, there are developers constructing housing that is affordable to people at 50–60% area median income, without subsidy.[1] These projects include microunits (very small single occupancy units) or co-housing in which residents live in a small unit with shared facilities such as community rooms or large kitchen and dining spaces. Beyond these private market alternatives, all other affordable – and deeply affordable – housing is constructed with the benefit of some type of subsidy. But, as mentioned above, even subsidized housing is often constructed by private market actors.

Conclusion

This chapter highlights all of the different players and actors that are involved in the development and operation of affordable housing. Government, at the federal, state, and local level interact with various private actors to create the affordable housing sector in the United States. No longer does the federal government play a dominant role as it once did during the public housing era of the mid-20th century. Rather, we have entered an era of public-private partnerships as the primary mechanism to create affordable housing. This shift has not occurred without opposition. Many people have credibly argued that the increasing reliance on the private market to deliver a basic need – such as housing – is like trying to jam a square peg in a round hole. Regardless of one's view on this particular debate, there is little dispute that the current approach to affordable housing is inadequate. The high levels of housing cost burden and homelessness in the United States speaks to the shortcomings of our current approach.

There is likely no example that fully highlights the interaction between multiple actors than the use of LIHTC to construct affordable housing. As you will see in the next several chapters, the LIHTC program combines many different players in one affordable housing project. In the United States, government-financed housing is not government developed or government managed housing. Without the seamless interaction of multiple players, housing of this type could not be constructed. There are many critics of the LIHTC program for this very reason. Many argue that it would be more efficient to have the government play a more significant role and eliminate the additional fees and complexities associated with the program. On the other hand, there are many people who argue that a government run and managed housing program would be inefficient and unsuccessful. The current model can best be described as a political compromise that came into existence following the demise of the public housing program in the United States.

Note

1 Refer to, for example, the Matter of Appeal fo Seattle Mobility Coalition www.google.com/Url?Sa=T&Source=Web&Rct=J&Opi=89978449&Url=Https://Web6.Seattle.Gov/Examiner/Case/Document/26477&Ved=2ahukewjiqafwh_2faxwrozqihdqddnuqfnoecbwqaq&Usg=Aovvaw3hfogsxb_X7hvgpxy_Ilrg

References

Abdelhadi, S., & Aurand, A. (2023). *State and local investments in rental housing: A summary of findings from the 2023 rental housing programs database*. National Low Income Housing Coalition. https://nlihc.org/sites/default/files/2023–10/state-and-local-investments-rental-housing-report.pdf

Anglin, R. V., & Montezemolo, S. C. (2004). "Supporting the community development movement: The achievements and challenges of intermediary organizations." In R. V. Anglin (Ed.), *Building the organizations that build communities. Strengthening the capacity of faith- and community-based development organizations*. Washington, DC: U.S. Department of Housing and Urban Development, Office of Policy Development and Research.

Bratt, R. G. (2008). Nonprofit and for-profit developers of subsidized rental housing: Comparative attributes and collaborative opportunities. *Housing Policy Debate, 19*(2), 323–365.

City of Seattle (2024). *Housing Levey Overview*. Seattle, WA. Retrieved from www.seattle.gov/housing/levy#:~:text=Since%201986%20the%20Housing%20Levy,homeownership%20opportunities%20throughout%20the%20city

Ellen, I. G., & Voicu, I. (2006). Nonprofit housing and neighborhood spillovers. *Journal of Policy Analysis and Management, 25*(1), 31–52.

Galster, G., Levy, D., Sawyer, N., Temkin, K., & Walker, C. (2005). *The impact of community development corporations on urban neighborhoods*. Washington, DC: The Urban Institute.

Glickman, N. J., & Servon, L. J. (2003). By the numbers: Measuring community development corporations' capacity. *Journal of Planning Education and Research, 22*, 240–256.

Joint Center for Housing Studies. (2023). *The state of the nation's housing*. Cambridge, MA: Joint Center for Housing Studies, Harvard University.

Keightley, M. (2022). *Ownership of the U.S. rental housing stock by investor type: In brief*. Washington, DC: Congressional Research Service.

Khadduri, J. (2015). "The founding and evolution of HUD: 50 years, 1965–2015." In L. M. Ross (Ed.), *HUD at 50 creating pathways to opportunity*. Washington, DC: U.S. Department of Housing and Urban Development Office of Policy Development and Research.

Kimura, D. (2019). The hybrid: A new deal design. *Affordable Housing Finance*. Retrieved from www.housingfinance.com/finance/the-hybrid-a-new-deal-design_o

Kleit, R. G., & Page, S. B. (2008). Public housing authorities under devolution. *Journal of the American Planning Association, 74*(1), 34–44.

Kleit, R. G., & Page, S. B. (2015). The changing role of public housing authorities in the affordable housing delivery system. *Housing Studies, 30*(4), 621–644.

New York City Housing Authority. (2023). *NYCHA 2023 Fact Sheet*. New York City: New York City Housing Authority.

Nguyen, M. T., Rohe, W. M., & Cowan, S. M. (2012). Entrenched hybridity in public housing agencies in the USA. *Housing Studies, 27*(4), 457–475.

O'Regan, K. M., & Quigley, J. M. (2000). Federal policy and the rise of nonprofit housing providers. *Journal of Housing Research, 11*(2), 297–317.

Quercia, R. G., & Galster, G. C. (1997). The challenges facing public housing authorities in a brave new world. *Housing Policy Debate, 8*(3), 535–569.

Scally, C. P. (2009). State housing finance agencies forty years later major or minor players in affordable housing? *Journal of Planning Education and Research, 29*, 194–212.

Scally, C. P., Curran-Groome, W., Kort, A., Kumari, S., & Lo, L. (2023). *The state of community-based development organizations: Results from the sixth national census of community-based development organizations*. Washington, DC: Urban Institute. Retrieved from https://naceda.memberclicks.net/assets/gv/The%20State%20of%20Community-Based%20Development%20Organizations.pdf

United States Department of Housing and Urban Development (HUD). (2022). *Picture of subsidized households*. Washington, DC: United States Department of Housing and Urban Development.

United States Interagency Council on Homelessness. (2024). *USICH, our mission, our values, our history*. Retrieved from www.usich.gov/about/usich

Walter, R. J., Colburn, G., Yerena, A., Pederson, M., Fyall, R., & Crowder, K. (2020). Constraints and opportunities for innovation in the moving to work demonstration program. *Housing and Society*, *47*(1), 1–21.

Chapter 8

Supply Side Housing Assistance

Overview of Supply Side Housing Assistance

Supply side housing encompasses all programs, initiatives, and strategies that aim to increase the supply of housing. Supply side programs stand in contrast to demand side assistance that provides benefits directly to households. Determining the appropriate allocation of housing programs and policies – between the supply and demand side – is a fundamental responsibility and challenge for policymakers. As described in Chapter 9, demand side housing support has become an increasingly important part of federal low-income housing policy over the last four decades. Historically, supply side programs, such as the public housing program, dominated federal housing policy in the United States. The public housing program is the most well-known and oldest supply side program, and its subsequent manifestations include HOPE VI, Choice Neighborhoods, and the Rental Assistance Demonstration (RAD) program. Other supply side programs provide direct subsidies to construct or rehabilitate housing such as tax credits, grants, or loans. Currently, the largest supply side production program in the United States is the Low-Income Housing Tax Credit (LIHTC), which was established in 1986 as part of President Ronald Reagan's tax reform legislation. It is common that other housing supports – such as demand side housing programs such as rental assistance – are used in combination to provide enhanced affordability.

This chapter begins by exploring the public housing program, how it operates, and the characteristics of the public housing stock and its tenants. We discuss the well-documented challenges of public housing and the decline of the program. Narratives about public housing are complex and conflicting. Scholarship highlights the significant success achieved by the public housing program, but high-profile failures such as Cabrini Green in Chicago and Pruitt Igoe in St. Louis have shaped the broader narrative (Goetz, 2013). In response to the perceived *failures* of public housing, the HOPE VI program, Choice Neighborhoods, and the Rental Assistance Demonstration Program were established. We describe these programs, their intended purposes, and the challenges they face. Special attention is given to the impact of certain failures of public housing on public perceptions of government involvement in the provision of affordable housing. This shift in perception (whether accurate or not) has had a profound impact on affordable housing policymaking over the last half century.

The launch of the LIHTC program ushered in a new era of supply side housing support. We explore the mechanisms of the tax credit program and how credits are allocated to developers of affordable housing. Given the cost of constructing housing, it can be difficult for developers to build affordable housing with only one subsidized financing source (the tax credit). Most affordable housing projects use several financing mechanisms. All levels of government (federal, state, and local) are involved in financing and structuring the development of affordable housing

DOI: 10.1201/9781003356585-11

with tax credits. We also know that there is increasing use of demand side rental assistance (vouchers, for example) in tax credit-financed housing projects which helps provide enhanced affordability (Colburn et al., 2024). And project-based rental assistance is often used to provide deeper subsidy in tax credit developments. But project-based rental assistance can also be used in range of housing products that are not supported with tax credits. Other supply-side funding includes federal block grants like the Community Development Block Grant Program or HOME Investment Partnerships Program, and a variety of state or local funds. Common sources of funding from states include housing trust funds and tax-exempt housing bonds. Examples of local supports include linkage fees and tax increment financing. We examine each of these funding mechanisms in detail throughout this chapter.

This chapter concludes with a discussion of the challenges of supply side housing assistance, such as the difficulty of providing housing to the lowest income households with a single subsidy and the challenges of constructing affordable housing with multiple subsidies and sources of funding that are frequently required. Unlike demand-side housing programs, some supply-side tenants tend to be higher income because the supply side subsidies like LIHTC are not sufficiently deep or robust to serve the lowest income households. Other challenges include the inability of the tax credit program to develop units in areas where there is the greatest need and its demonstrated inability to provide adequate locational choice to tenants.

Federal Housing Development Programs

The public housing program, despite no longer being the largest housing program, is among the most well-known housing policies in the United States. Since the inception of the public housing program in 1937, the government constructed over 1.3 million housing units (Goetz, 2011). These units represented an important addition and complement to privately-financed housing in the United States. Since the 1980s, the public housing stock has been dwindling due to funding cuts that has led to disinvestment and a shift in the focus of U.S. housing policy. In response to the declining quality of the public housing stock, a range of other programs have emerged to help redevelop and transform public housing, including HOPE VI, Choice Neighborhoods, and now the Rental Assistance Demonstration program.

Public Housing

The public housing program was established in 1937 as part of the New Deal policymaking in response to the Great Depression. Public housing has long been the most well-known and most significant supply-side program that has housed millions of low-income households in the United States. Currently, there are just under 1 million public housing units remaining that house approximately 1.8 million people (CBPP, 2021). Residents of public housing are disproportionately disabled (22%), older adults (18%), children (37%), and reside in female-headed households (62%) (CBPP, 2021). Tenants must earn under 80% of the area median income to qualify for public housing although many households that live in public housing are below 30% of the area median income because the program prioritizes extremely low-income households.

Public housing authorities (PHAs) developed the public housing stock in the United States and continue to assume responsibility for the operation and maintenance of the existing stock of housing. When they built public housing, PHAs used funds raised by issuing government-backed tax-exempt bonds and the funds were directed to PHAs pursuant to an Annual Contributions Contract (ACCs). Once the public housing was constructed, there was an ongoing need for operating funds to support the management and upkeep of these housing units. During the first few decades of the public housing program, maintenance and property upgrades were primarily covered by the rent collected from tenants. In the early years, public housing didn't serve the lowest income households, rather it served moderate income households who contributed a portion of their monthly income toward rent which provided an ongoing source of funds for building operations. Over time, the public housing program began to serve lower income households. The motivation for doing so was logical – we should allocate public housing units to the households with the greatest needs. While noble, this decision radically reduced the rental income received from tenants which limited the funds available for operations and maintenance.

As rent collections fell in public housing projects, there was increased pressure to collect more rental income from tenants. In response, the Brooke Amendment was passed in 1969 to protect public housing tenants and capped the portion of a tenant's income that must be contributed toward rent at 25%. This threshold was later increased to 30% in the 1980s. Today the standard rent contribution by tenants is approximately 30% although some households may pay up to 40%. Although well-intended, the amendment had a dire consequence for PHAs as the reduced cash flow jeopardized their ability to cover the costs of maintaining the aging stock of public housing. The program was further weakened in the 1970s, when President Nixon issued a moratorium on funding of new public housing development.

The shortfall in operating cash flow (described above) caused HUD to begin supporting PHAs with the Public Housing Operating Fund for operating costs and the Public Housing Capital Fund for major renovations and replacement of essential items like heating and cooling equipment. Even with the two additional funding streams, public housing remained underfunded which led to an increasingly poor-quality public housing stock in need of major renovation. The halt on new construction and lack of funding for the upkeep of existing properties led to the creation of new programs like Section 8 rental subsidies.

Beyond limited operating funds and the halt in new construction, public housing also faced other challenges. At its inception, private developers opposed the creation of a public housing program due to concerns that it would be competitive and jeopardize the profitability of private housing development and management. To appease private interests, public housing was constructed in a way to avoid it being perceived as competitive. The primary way in which this was achieved was through the relatively low quality of this housing. The subpar design and construction contributed to the high maintenance costs that were required in the decades after construction. Additionally, local political opposition to public housing had significant impacts on the siting of these housing developments. Many public housing projects were often confined to low-income, high-crime areas, which compounded the challenges of maintaining and managing public housing developments. Pruitt-Igoe in St. Louis, Missouri stands as a prime example of a large public housing project located in a low-income, high-crime area of the city. Because of budget cuts during construction, inadequate funding for upkeep and security, and ineffective management, the development was torn down less than twenty years after it was built. For opponents of public housing, Pruitt-Igoe is frequently cited as an example of the failure of the public housing program in the United States.

While the failures of public housing have been well-publicized, there is a counter argument that suggests that a few high-profile failures have masked the significant successes of the program (Goetz 2011). Even in the face of inadequate funding for both construction and maintenance, the public housing program has served millions of low-income households over the last century. In many cities, including New York, the public housing stock remains a vital component of the housing supply and provides much needed affordability. While public attention has focused on notorious projects such as Pruitt-Igoe in St. Louis and Cabrini-Green in Chicago, public housing continues to quietly serve millions of people without the general public even knowing that many units in their city are owned and operated by local PHAs.

As frustration with the public housing program grew and the quality of the public housing declined, major changes were implemented beginning in the 1990s. Over the last thirty years, hundreds of thousands of units have been torn down, replaced with alternative funding methods like Low-Income Housing Tax Credits, or converted to Project-Based Vouchers or Project-Based Rental Assistance through the Rental Assistance Demonstration. PHAs have faced criticism for replacing public housing units, as the housing that has replaced it tends to cater to higher income households, leaving fewer options for extremely low-income households in need of deep subsidies. For example, many former public housing residents may not be able to afford to live in tax credit developments without the additional support of a Housing Choice Voucher. Another criticism is that the PHAs have lost many supported units of housing as large public housing developments have been replaced with smaller mixed income developments that serve far fewer households. Given limited funding from the federal government, PHAs have struggled to continue to provide high quality, affordable housing for as many households as possible. Each time a public housing project is renovated or demolished, the PHA must determine the best financing and subsidy model to use, how to support existing tenants through the redevelopment process, and what income levels can be served in the new development. The inadequacy of federal supports makes these decisions very challenging for local PHAs.

HOPE VI

The HOPE VI program was established in 1993 to transform public housing in the United States. The federal government had decided to no longer construct new public housing and faced significant challenges with the existing stock of public housing given declining quality and expensive maintenance. HOPE VI sought to redevelop public housing and address many of the well-documented issues that plagued distressed public housing through improved design and economic integration by promoting the development of mixed-income communities. The program was created by HUD to respond to the recommendations of the National Commission on Severely Distressed Public Housing. HOPE VI developments used mixed-financing structures often through public-private partnerships. The units were built at lower densities with amenities to mirror market-rate housing.

Funding for HOPE VI was discontinued after 2010 when Choice Neighborhoods was launched in its place. HOPE VI provided $6.7 billion during its tenure in the form of a series of grants for planning, revitalization, and demolition activities. Revitalization grants represented $6.2 billion of the total budget authority of HOPE VI. Under the program, 262 revitalization grants were provided to 133 housing authorities (HUD, 2010). With the support of these grants, a total of 97,389 new mixed-income units were produced. Over half were public housing units, another 30% were offered below market rate, and 13% of the units were market-rate (Gress et al., 2016). Some households were displaced from the original public housing site given that HOPE VI did not develop new housing at a one-to-one ratio. From 1995 through 2009, approximately 140,000 public housing units were lost which reduced the stock of public housing from 1.33 million to 1.19 million units. During this time, the number of vouchers grew, partially as a result of additional vouchers given to former public housing residents displaced by HOPE VI redevelopment (Turner & Kingsley, 2008).

The efficacy and wisdom of HOPE VI is contested, as the program has been found to have both positive and negative attributes. An oversimplified summary of HOPE VI is that for those households that were fortunate enough to reside in the newly renovated housing units, they reported high levels of satisfaction given the newer and higher quality housing and improved surroundings (Joice, 2017). Research following public housing residents after relocation has shown minimal improvements in employment, financial stability, and self-sufficiency. Less than one-third of the original residents returned to the new HOPE VI developments. While feelings of safety and security improved and there were some improvements in mental health, no other health benefits were observed. Educational and school outcomes showed minimal improvements (Gress et al., 2016).

Choice Neighborhoods

In response to the shortcomings of the HOPE VI program, HUD established the Choice Neighborhoods public housing demolition and redevelopment program as a successor program that incorporated many of the core principles of HOPE VI. The benefit of Choice Neighborhoods was that it took an additional step to not only redevelop a single property but the entire community. Launched in 2010, the key aim of the program was to redevelop distressed public housing projects with a one-for-one replacement of all public housing units. One-for-one replacement was not part of the HOPE VI program and, as described above, resulted in households being displaced. Choice Neighborhoods requires a community Transformation Plan that focuses on housing, people, and neighborhoods. The program uses a diverse portfolio of public and private funding to create communities with diverse housing types and incomes.

Under Choice Neighborhoods, PHAs must take a community-based approach and collaborate with public agencies, non-profits, and the private sector. This program emphasizes the importance of supportive services to assist residents. Enhancing quality of life for residents is an essential focus of the program. To deliver these results, Choice Neighborhoods developments create amenities such as recreation areas and other places in which the community can gather and interact. The program also prioritizes education through partnerships with schools, non-profits, and community groups to expand educational opportunities for residents in the community. The focus on education is a direct response to the cycle of poverty that was evident in many public housing projects over time.

Over the last decade, Choice Neighborhoods has allocated over $1 billion to PHAs in 40 cities to create 40,000 new mixed-income units (National League of Cities, 2022). Although the program is still in place, it comes with a substantial price tag and there isn't sufficient funding to redevelop all the public housing projects that each PHAs would like to. Furthermore, the resources required to transform an entire community are extensive, therefore, the program relies heavily on local PHAs to leverage significant additional public and private dollars that may, or may not, be available and accessible. Smaller PHAs often do not have the organizational capacity to even apply for a Choice Neighborhoods grant, let alone implement it. Despite the constraints that have limited the scope and scale of the program, when implemented, the outcomes of Choice Neighborhoods have been quite positive. Besides the improvement of dilapidated housing, the program has spurred economic development and attracted new investment to under-resourced communities, built strong social networks, reduced crime and made communities safer, and integrated community services such as educational and employment opportunities for residents (Joice, 2017).

Rental Assistance Demonstration (RAD) Program

Authorized in 2012, the Rental Assistance Demonstration (RAD) program is one of HUD's newest programs that is designed to address the capital needs of public housing. According to HUD, RAD was specifically created to address the $26 billion-dollar public housing deferred maintenance needs. RAD allows PHAs to convert public housing into project-based vouchers (PBVs) or project-based rental assistance (PBRA). The purpose is to establish reliable funding sources that facilitate borrowing for PHAs and enables them to utilize financing options like LIHTC for property upgrades. Residents of these units have the right to return to the rehabilitated public housing properties without any additional screening or requirements. This is different than programs like HOPE VI which displaced residents during rehabilitation and had new requirements to return to the original site after redevelopment. In addition to public housing, RAD allows owners of privately owned housing units that participate in HUD's legacy programs (Rent Supplement, Rental Assistance Payment, and Section 8 Moderate Rehabilitation) to convert units to new PBV or PBRA contracts. RAD can be used in conjunction with other programs like Choice Neighborhoods or LIHTC.

RAD provides a stable operating subsidy that can be leveraged to raise private capital for redevelopment. Private investors partner with building owners during the redevelopment process since there is an opportunity to receive a return on their investment from tax credits and rental income. The original cap for the number of units that could be converted was 60,000. The current cap is 455,000 units nationwide (Treskon et al., 2022). However, the program has operated below the allowable cap since it takes roughly 2.5 years after receiving approval to convert units. In 2022, 163,973 public housing units were converted, 54,122 units had approval to be converted, and another 124,467 units were slated for conversion in the future. In addition to the public housing program, there are 42,907 HUD-assisted housing units owned by private owners that have already been converted using RAD. Currently, 2,165 units are in the process of conversion, while another 11,516 units are scheduled for future conversion (Gramlich, 2023).

Project-Based Rental Assistance

There is a class of rental supports that are attached to specific housing units. In general, these programs are known as project-based rental assistance. It is an interesting question as to whether these supports are a supply-side or demand-side housing program. The rental assistance operates similarly to a voucher or rental assistance in that it provides rental support to a tenant in that particular unit. So, in that sense, one could think of these programs as a demand side benefit. But because these subsidies are attached to specific housing units, one can also make an argument that they are a supply side benefit. Ultimately, the categorization doesn't really matter. But we highlight these programs in this section, because the presence of project-based rental assistance can promote the development of new housing in some circumstances and in all other instances where new housing is not built, the unit is preserved for a certain period as affordable. While project-based rental assistance does not always increase housing supply, there is at least an increase in the number of affordable units available.

HUD manages a range of supply-side rental assistance programs called Project-Based Rental Assistance (PBRA). The umbrella of PBRA includes a variety of different programs including the Rental Supplement Program (Rent Supp), Section 8 New Construction and Substantial Rehabilitation Program (S8 NC/SR), Rental Assistance Payments (RAP), and Project Rental Assistance Contracts (PRAC/202 and PRAC/811). All these programs are designed to provide rental support to low-income households by attaching a subsidy to a specific housing unit. Similar to demand-side rental assistance, the subsidy covers the difference between what the tenant pays toward rent (typically 30% of household income) and the contract rent. Property owners with PBRA contracts agree to rent the units (with PBRA attached to them) to low-income households and adhere to federal rules including housing quality inspections. Most PBRA contracts are 20 years with the option to renew. The PBRA programs are used in tandem with many other affordable housing programs including LIHTC and RAD as described above.

Other PBRA examples include smaller or more targeted rental assistance programs like Rent Supp, PRAC/202 for older adults, and PRAC/811 for people with disabilities. PBRA programs serve approximately 1.2 million households throughout the country (HUD, 2024). Demand for these programs far exceeds the annual discretionary appropriation that funds the programs. HUD has not had the authority to enter new contracts since the 1980s due to the launch of RAD in 2012. Over time, PBRA contracts have become more expensive due to rising rents that increase the cost of rental support (Congressional Research Service, 2023).

Federal Multifamily Housing Programs

There are other federal programs that specifically support the financing of affordable housing owned by the private sector and nonprofits that fall within HUD's Office of Multifamily Housing. These programs provide favorable financing options and mortgage insurance through the Federal Housing Administration to expand and preserve the affordable rental housing stock. Favorable financing options include capital advances and lower interest rates that reduce the cost of development and debt service during operation. By reducing risk for lenders, the mortgage insurance programs make it easier for affordable housing developers to obtain funding for new construction and rehabilitation. Some of these programs also use project-based rental assistance as described above.

For example, Section 202 – Supportive Housing for the Elderly, which was established in the Housing Act of 1959, is also listed under PBRA because the program receives project-based rental assistance for operations but also capital advances for construction and rehabilitation. The focus of the program is to increase the supply of housing for people over the age of 62 with incomes below 50% of AMI. Funding is provided to nonprofit organizations to develop and operate housing for older adults in the form of a capital advance for the construction, rehabilitation, or acquisition of housing or operating funds. The capital advance does not accrue interest and does not need to be paid as long as the development stays affordable for very low-income older adults for a minimum of 40 years. This program alone has served over 400,000 households in its history. The average annual household income of Section 202 tenants is $13,300 and the average age is 79 (Couch, 2019).

Other active multifamily HUD housing programs that fall within this category are Mark-to-Market Program (M2M), Mortgage Insurance for Cooperative Housing (Section 213), Mortgage Insurance for Purchase or Refinance of Existing Healthcare and Multifamily Rental Housing (Sections 207 and 2239f), Mortgage Insurance for Rental Housing for the Elderly (Section 231), Mortgage Insurance for Rental Housing for Urban Renewal and Concentrated Development Areas (Section 220), Mortgage Insurance for Supplemental Loans for Multifamily and Healthcare Projects (Section 241), Multifamily Mortgage Risk-Sharing Programs (Section 542(b) and 5429c)), and Multifamily Rental housing for Moderate-Income Families (Section 221(d)(3) and (4)). Renewal of Section 8 Project-Based Rental Assistance and Supportive Housing for Persons with Disabilities (Section 811) also fall within HUD's Office of Multifamily Housing along with Supportive Housing for the Elderly Program (Section 202).

National Housing Trust Fund

The National Housing Trust Fund was established under the Housing and Economic Recovery Act of 2008. The first $174 million in funding was distributed to states in 2016. In subsequent years, additional funding has been allocated each year, and in 2023, $382 million was designated for the fund. The National Housing Trust Fund is the first new source of dedicated funding for housing since 1974. The funds are used to support the lowest income households with acute housing needs. First-time homeowners may use up to 10% of the funds, with the remaining resources designated for affordable rental housing production, preservation, rehabilitation, or operation (National Low Income Housing Coalition, 2024). Unlike most other programs that serve households up to 80% of AMI, the National Housing Trust Fund, at a minimum must allocate 75% of funds to extremely low-income households below 30% of AMI. Additionally, all funds must serve households with very low incomes earning no more than 50% of AMI. The program operates as a block grant, allowing each state to customize its use of funds as it sees fit.

Low-Income Housing Tax Credit Program

The Low-Income Housing Tax Credit (LIHTC) Program originated in 1986 and was permanently established in 1993. Given that the U.S. no longer develops public housing, LIHTC is now the primary supply-side affordable housing production program in the United States. Because the benefit is a tax credit, not an outlay of cash, the program is administered by the IRS not HUD. LIHTC, therefore, does not share the same bureaucratic home as most of the other major housing programs described in this book. The general concept of the program is the government provides tax credits to developers of affordable housing (both non-profit and for-profit developers, PHAs, or other public-private partnerships) to construct housing with specified affordability requirements. State housing finance agencies award tax credits to developers as directed in their state Qualified Allocation Plans (QAPs). Developers submit proposals based on the state criteria and each proposal is assessed on the factors outlined in the QAP. Preference in the scoring criteria is given to developers that have the longest affordability periods and serve the lowest income households. Affordability periods are at least 30 years but are often much longer to increase the chances of receiving a tax credit in the competitive process. At the end of the compliance period, which is typically 15 years, the owner of the property will either refinance the development or sell it. Although there was concern in the past about expiring LIHTC properties converting to market rate, the majority of LIHTC properties have remained affordable.

Developers, in essence, use tax credits to raise the money that serves as the equity capital for a housing development. The developers are granted tax credits by the state housing finance agency and the developers then sell the tax credits to investors (such as banks and insurance companies) that use the tax credits to offset or reduce their tax liability. Investors may also receive other benefits such as pass through losses and depreciation expense that may reduce their tax liability, surplus cash flow, and residual value of the project if it is sold. The developers then use the proceeds from the sale of the tax credit to finance the construction of the housing project. The development is structured through a partnership agreement and investors are limited partners and the general partner (the developer) manages the day-to-day operations. There are two types of tax credits: the 9% credit and the 4% credit. The 9% credit is often used for new construction and provides about 60–70% of equity into the project. The 4% credit is commonly used for rehabilitation with bond financing and provides about 30% equity for the project. In both cases, the tax credit alone does not finance the entire project, therefore the developer must fund additional sources of funding to complete the financing. The 9% credit is provided to states by the IRS based either on a per capita basis or a minimum floor amount. Given the smaller size of a 4% tax credit, tax-exempt bonds are issued in connection with the issuance of the tax credit. Therefore, the availability of bonding capacity at the state level frequently determines the scale of the 4% tax credit program in each state. The price that investors pay for tax credits fluctuates with market conditions. Prevailing corporate tax rates and the level of interest rates may impact the pricing of tax credits in the market. Over the last ten years, the prices for tax credits have traded between $.87 and $1.05 per tax credit dollar (Novogradac, 2024). When the price of tax credits is low, developers get less money for the tax credits which requires them to find additional funding to fill the funding gap needed to develop the property.

Since inception, the LIHTC program has produced over 3 million affordable housing units. Given data constraints associated with the LIHTC program, it is difficult to capture a full picture of tenant characteristics for residents who live in LIHTC units. However, compliance data reported to HUD by some states provides a general understanding of who resides in LIHTC units. Approximately half of LIHTC tenants are extremely low-income, earning less than 30% of the area median income. The majority of the other tenants fall in the low-income category (30 to 50% of area median income). Since the LIHTC program does not provide a deep subsidy, approximately one third of LIHTC tenants use another subsidy to access and afford LIHTC housing and one third are cost burdened spending more than 30% of their income on rent (Scally et al., 2018). As rental housing costs have risen over time, the share of tenant-based rental assistance used in LIHTC units doubled (Colburn et al., 2024).

Since the federal government ended its investment in the development of public housing, LIHTC has become the primary mechanism to build affordable housing in the United States. The three million units constructed with the support of tax credits speaks to the importance of the program. Even among those who take issue with the structure and complexity of the program, the end result is a robust stock of housing that provides affordability to millions of households. Despite the significant production achieved under this program, there are notable challenges and shortcomings. First, LIHTC units are not required to remain affordable beyond the initial affordability period which is typically 30 years. A second critique is that LIHTC doesn't provide deep subsidies like the public housing program so frequently very low-income tenants use multiple subsidies (LIHTC plus rental assistance) to access LIHTC units (Colburn et al., 2024). A third critique is the complexity of the program, which creates significant legal and transaction fees for many parties that help to arrange and structure these deals (Keightley, 2017).

Finally, the location of LIHTC developments is a concern. In an echo of the challenges of siting public housing projects, the structure of the LIHTC program combined with frequent community opposition has caused LIHTC developments to be disproportionately located in communities that have been subject to disinvestment. LIHTC properties, although more geographically dispersed than other affordable housing developments, tend to be in high-poverty, racially concentrated neighborhoods (Ellen et al., 2016). A landmark case highlighted this issue. In 2015, Inclusive Communities, a Texas-based non-profit organization, claimed that 92.29% of LIHTC units in Dallas, TX were located in census tracts with less than 50% white residents (*Texas Dept. of Housing and Community Affairs v. Inclusive Communities Project, Inc.*, 2015). The non-profit sued the Texas Department of Housing and Community Affairs for perpetuating residential segregation by disproportionately allocating tax credits to minority communities. The Supreme Court ruled in favor of Inclusive Communities Project, Inc., holding that disparate impact—unintentional discrimination that disproportionately impacts a group of individuals—violates the Fair Housing Act. This decision was significant in promoting fair housing in the United States and provided an opening for more legal challenges based on disparate impact arguments. Many housing finance agencies have since incorporated provisions in their qualified allocation plans that support tax credit development in low poverty areas.

Block Grants

Block grants are another tool used by the federal government to provide money for affordable housing. The federal government uses block grants to allocate funding to states and cities for local initiatives. The difference between a block grant and funding for a specific program (such as Housing Choice Vouchers) lies in the level of flexibility in its use. Supporting the development of affordable housing is one of many community development activities for which block grants funds may be used. Block grants are useful to fill financing gaps for affordable housing development, but they do not serve as a primary financing source for development projects (unlike LIHTC which does serve that function). Block grant programs are large; there are 22 different types of block grants that distributed $59.1 billion to state and local governments in 2022 based on formulas established by each program. Although some of these block grants can be used for housing and other community development and social services activities, others are specifically for public health, transportation, infrastructure, or law enforcement (Congressional Research Service, 2022).

Supporters of block grants argue that state and local governments are better equipped to address the needs of their residents than is the federal government. They also argue that local control leads to greater efficiency and cost-effectiveness and promotes long-term planning with stable funding. However, opponents argue that block grant funding may not always be directed to the areas of greatest need and fear that public support for funding may diminish over time due to the lack of a clear, collective impact given the disparate programs that they support and the difficulty in assessing program outcomes.

The Community Development Block Grant (CDBG) is most commonly used to fund and support affordable housing. This program was established in the Housing & Community Development Act of 1974. It combines eight competitive categorical grants into a single grant to allow states and local governments to flexibly address housing needs. The main objectives of the program are to provide affordable housing, economic opportunities, and public services and infrastructure for low- and moderate-income households in lower income communities. 70% of the funds are distributed to entitlement communities (municipalities with over 50,000 residents) while the remaining funds are distributed to states to allocate to smaller communities that do not receive entitlement funds. Although the funds can be used for a wide range of activities, there are certain uses that are prohibited such as government buildings, political activities, and new housing construction. Allowable uses of CDBG funds for housing include rehabilitation, homeownership and rental assistance, and energy efficiency upgrades. Related activities such as comprehensive planning, housing planning, and fair housing analyses are also frequent uses. Some jurisdictions allocate a significant portion of their CDBG funds toward affordable housing, while others only allocate a minimal amount.

The HOME Investment Partnerships Program is another block grant that provides funds specifically to support affordable housing for low- and moderate-income households. Eligible uses include purchasing existing housing or vacant land, constructing and rehabilitating housing, demolishing structures, relocation support, fair housing services, and soft costs incurred in the development of affordable housing development such as engineering plans and attorney fees.

60% of HOME funding is awarded to local governments and the remaining 40% is allocated to states to distribute to non-entitled cities and towns. Recipients of HOME funds must match 25 cents for every dollar awarded using state, local, or private contributions through cash, bond financing, materials, labor, or donated property.

The Indian Housing Block Grant (IHBG) is another block grant that provides funds specifically for housing. IHBG is the single largest source of Native American housing assistance. Funds may be used for new housing construction and rehabilitation, and housing programs and services. Recipients of IHBG funds include federally recognized Tribes, Tribally Designated Housing Entities (TDHEs), and a limited number of state-recognized Tribes. While CDBG and HOME are entitlement grants (they are automatically distributed to communities), IHBG funds are allocated to eligible recipients via a competitive process. Allocation decisions are made annually based on a range of factors, including, the need for funding, capacity of the applicant, soundness of approach, and comprehensiveness and coordination of the proposal.

State and Local Financing for Affordable Housing Development

Because federal programs do not provide sufficient funds to meet local needs, state and local governments have their own programs to supplement the dollars received from Washington, DC. These programs include state or local housing trust funds, tax-exempt housing bonds, linkage fees, and tax incentive programs to fund affordable housing such as state housing tax credits.

Housing Trust Funds

A housing trust fund is a pool of money that can be used to support affordable housing development and other affordable housing activities and supports. The fund distributes money on to recipients, such as cities, counties, nonprofit organizations, and developers. The ability of a housing trust fund to make a meaningful impact is dependent on the size of the fund. Smaller housing trust funds distribute fewer funds, and by extension, have a more limited impact.

Housing trust funds – first established in the 1960s – became a popular tool to support affordable housing in the 1980s (Connerly, 1990). States like California and New York were two of the first states to experience pronounced housing supply shortages, rising housing costs, severe rent burdens, and homelessness at the time of waning federal support for affordable housing (Brooks, 2007). Those states established housing trust funds to help address these shortfalls and, today, they are widely used across the United States to generate funding to support the development of affordable housing and related activities. Housing trust funds are created by state and local governments through enabling legislation and a division within government usually manages these funds and operate the housing trust fund.

The funding mechanisms for housing trust funds vary. Many housing trust funds are established with a dedicated source of funding, although this is not a requirement. When housing trust funds do not have a specified source of funding, they are financed through allocations from the state general fund. A dedicated source of perpetual funding is needed for stable and continuous distribution of money from the trust fund. The most common form of a dedicated funding source is a real estate transfer tax that funds the housing trust fund on a continuous basis. Other funding sources include document recording fees, revenues from state housing finance agencies, interest from real estate escrow accounts, bond proceeds, contractor's excise tax, foreclosure filings, interest on title escrow accounts, state income tax, tobacco tax, unclaimed property funds, and utility charges (Center for Community Change, 2016). Housing trust funds with dedicated funding sources tend to be more successful and have a greater impact over the long run.

Funding from housing trust funds is typically distributed through a competitive process as grants or loans. Eligibility to receive funding varies on the trust but funds usually serve low-income households, which are households that earn below 80% of the area median income. Funds are used for a variety of housing related activities including the creation and preservation of affordable housing units, supportive housing and homeless prevention, home improvement programs, down payment assistance, foreclosure prevention, programmatic support, capacity building, and technical assistance.

Funding commitments by housing trust funds are frequently not sufficient to support new construction or rehabilitation activities so multiple funding streams are used in conjunction with trust funds. A major critique of housing trust funds is that they don't provide sufficiently deep subsides and therefore cannot support households with the greatest needs (Larsen, 2004). This is true for many state and local programs, they only provide shallow subsidies that often serve higher-income households (Mueller & Schwartz, 2008). But even if funds are insufficient to support housing construction, trust funds can still make meaningful contributions by supporting programs or initiatives such as foreclosure prevention, down payment assistance, home education and counseling, or rental subsides.

Tax-Exempt Housing Bonds

Municipal bond financing is a mechanism governments use when they need to raise revenue for projects that serve the public good. Affordable housing projects fall within this purview. Municipal bonds are debt securities issued by governments to investors. To provide municipal bond issuers with an advantage in the capital markets, the government has made municipal bonds exempt from federal taxation. Therefore, investors benefit from owning municipal bonds because they do not pay tax on interest earned while owning those bonds. The end result is that municipal issuers don't need to pay as high an interest rate when they issue bonds which saves the municipality money. For example, if an investor pays 30% in federal taxes, that investor is indifferent between a 6% taxable bond and a 4.2% tax-exempt municipal bond. (After the investor pays 30% tax on the 6% interest that they earned, they are left with a net 4.2% return.) Municipal bonds are frequently issued by states, cities, counties, or housing authorities to finance a range of projects including infrastructure and housing.

There is a particular use of municipal bonds that is relevant to our discussion of affordable housing. Bonds issued for the purpose of funding the development of multifamily housing are known as Housing Bonds, part of a category of bonds known as Private Activity Bonds. Each state receives an allotment of bond capacity from the federal government and that capacity is allocated across eligible projects in that state, typically by the state housing finance agency. Private Activity Bonds cover a range of activities (in addition to housing), but recently housing has consumed most of the cap. The bonding cap is set by the IRS Code based on a per capita formula. These bonds are issued alongside 4% tax credits to fund affordable housing projects in the state. Developers of 4% projects receive both a tax credit and an allocation of bonding to finance the majority of the project. The 9% tax credits are larger and do not come with any bond financing. Because 4% tax credits are not capped (unlike the 9% program which is limited at the state level) the only constraint on the size of this program is the bond cap. In the past, the demand for Private Activity Bonds has been less than their supply, but that has changed in recent years in many states. As a result, the state bond cap now constrains the amount of 4% tax credit deals that can be developed in many states.

LIHTC developers partner with an entity (typically a city, county, or housing authority) that petitions the state housing finance agency to issue a portion of the state's volume cap of Private Activity Bonds. The majority (at least 50%) of the development costs must be funded by bonds during construction. Besides construction, the tax-exempt bonds can be used for permanent financing as well but does not need to remain as a permanent source of financing. However, when bonds are used for construction and permanent financing, it is difficult for the project's income to cover debt service so additional subsidies are needed. Therefore, a hybrid approach is often employed to use bonds during construction and through permanent conversion to secure other sources (Biber, 2007). Regardless of the approach, the lower interest rates that come with bond issuance lower the debt service for the owner/developer which promotes housing affordability.

The LIHTC program is very complex and a detailed review of the program, how it works, the disparate funding sources, and a detailed review of program outcomes is beyond the scope of this book. For readers interested in a deeper dive on the topic, we include a number of valuable references in a footnote to this paragraph. The Urban Land Institute also has an extensive list of case studies that are accessible to members of that organization.[1]

Tax Incentive Programs

There are a number of tax incentive programs that fund affordable housing. These tax benefits make it easier for developers to offer units below market rate because the incentives bring down the total costs of development which allows operators to reduce rental rates for lower income tenants. Examples include property tax levies, real estate excise taxes, and multi-family tax exemptions. One example of a tax incentive program is Washington State's multi-family tax exemption. Under this program, developers receive a property tax exemption on the value of new construction or substantial rehabilitation of multi-family housing. Developers that take advantage of the multi-family tax exemption must adhere to specific affordability periods and set aside a percentage of units for low-income households. They must also verify that these households are income qualifying under the regulations. The tax exemption that the developer receives depends on the percentage of units they make affordable and the income levels of the tenants in those units.

States have also created state housing tax credits, which have gained popularity recently. The idea of these state programs is to supplement the federal tax credit program. In 2023, the state of Kansas began to issue a state tax credit in conjunction with federal 4% and 9% tax credit programs to provide additional funding for LIHTC projects. Minnesota has its own tax credit program funded entirely by taxpayer contributions to finance the development and preservation of housing. Approximately half of the states and Washington, DC have implemented some form of state housing tax credit (Affordable Housing Finance, 2023).

Linkage Fees

Linkage fees are used by local government to attach approval for nonresidential development to the development of affordable housing. The logic behind this approach is that nonresidential development increases the demand for affordable housing. This is a proactive strategy to mitigate potential future housing market shortfalls that arise due to additional nonresidential development. Linkage fees are enforced through the permitting process. Developers must choose between building affordable housing units, contributing a fee to a fund dedicated to affordable housing, or donating land for housing development. The specifics of these programs depend on factors such as the required number of affordable units, whether linkage fees are limited to certain areas like downtown or apply citywide, and whether participation in the program is mandatory or voluntary. Voluntary programs offer a range of incentives for developers, such as density bonuses or relaxed parking or setback requirements, to entice developers to participate.

CRAs and TIF

Community redevelopment agencies, also known as CRAs, are local governmental entities that are specifically created to pursue community and economic development initiatives in under-resourced neighborhoods. CRAs can implement any program or project that has been outlined in the CRA plan that improves the built environment or socioeconomic conditions of the area. The plan is established at the time the CRA is created. A CRA may develop affordable housing or establish programming to support housing stability for lower income residents. A CRA is created to address a specific need in a designated area that is identified in an assessment conducted by a local jurisdiction. The boundary of the designated area and entity are established through a local ordinance and is created for a specific period of time, frequently 20 years. A state statute is needed to formally create the geographic boundaries for a CRA. At formation, a plan is established for the CRA which outlines the range of initiatives and programs the entity will pursue and the CRA is then staffed to carry out the plan.

CRAs are financed through tax increment financing, also referred to as TIF. This tool freezes property taxes for the designated area at the time the CRA is established. From that point forward, the increase in property tax revenue generated from the designated redevelopment area, which is established in the enabling legislation for the CRA, is used to fund the activities of the CRA. The money invested in the designated area can increase the value of property in the area due to improvements which can then further increase tax revenue captured by the CRA. These additional funds can be used for further investment in the area. This virtuous cycle of investment is the hope and promise of CRAs. The actual community investments may be completed on a pay-as-you-go basis from TIF or the CRA can raise bond financing and private investment that are then later paid with proceeds from the TIF. Both California and Florida are two states that use CRAs extensively, especially for affordable housing activities. CRAs and TIF have been instrumental in leveraging public and private funding to increase the affordable housing stock and improve neighborhood conditions.

Conclusion

Research from Fannie Mae and other agencies suggests a deficit of between 3–7 million housing units in the United States. The National Low Income Housing Coalition estimates a deficit of 7.3 million housing units that are affordable and available to extremely low-income households (households with incomes below 30% of the area median income) across the nation. While a precise estimate of the housing shortage is difficult to pinpoint, it is clear that the U.S. faces significant affordable housing supply challenges. In this book, our concern is most directly focused on the shortage of affordable housing, but the broader shortage also contributes to affordability challenges. Research highlights that new housing development – even market rate development – can alleviate pressures in the housing market and promote greater affordability. But it is important to stress that, given the scale of the affordability challenges, building market rate housing alone will not solve the affordable housing crisis in the United States. Among the many associated challenges, one of the major issues is that the operating costs of housing frequently

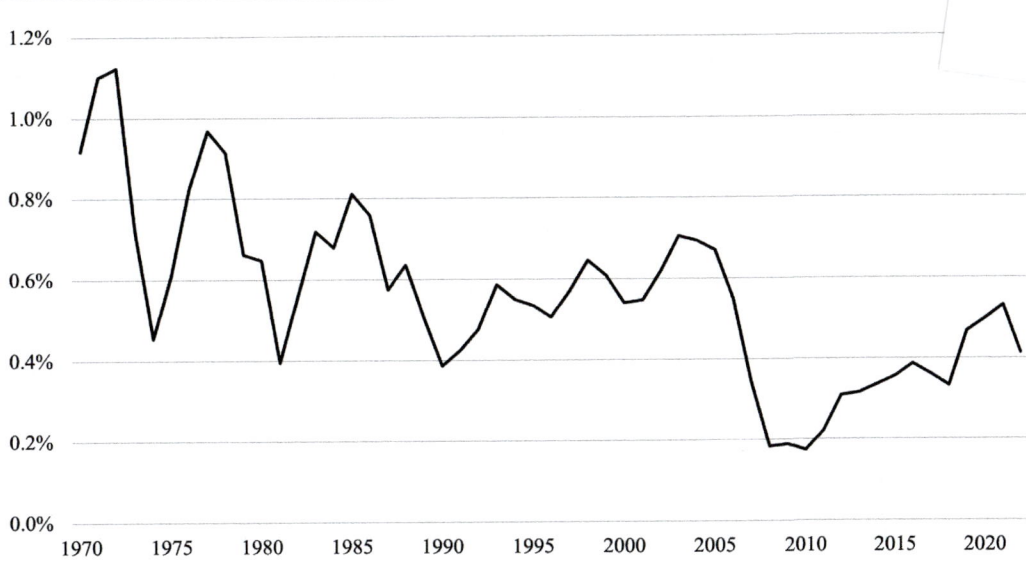

Figure 8.1 U.S. Housing Starts as a Share of the Population
(Source: U.S. Census Bureau)

exceed the rent that an extremely low-income renter could pay. Developers won't – without additional support – construct housing that rents are inadequate to support. The programs and policies described in both this chapter and the demand-side programs described in Chapter 9 are necessary to create the access to affordable housing that is needed.

The answer to the important question of why this shortage of housing exists is complicated, but the consequences are clear. The shortage of housing has produced both higher home prices and rents. While these increases are a problem for homebuyers and renters alike, one would assume that these market dynamics would be a strong incentive for developers to construct far more housing units, but the reality is that has not happened. Evidence of this underproduction is evident in Figure 8.1 that depicts the fall in per capita housing starts over time. Per capita housing development has been on a sustained decline over the last 50 years. The nation's affordability crisis is one unfortunate byproduct of this underproduction.

The reasons for the housing shortage are many. Opposition to housing development – NIMBYism – and constrained subsidy funding contributes to the problem. In addition, high construction costs, lengthy permitting processes, restrictive zoning regulations, limited land availability, associated infrastructure needs, and the need for additional utilities all make it challenging to develop affordable housing. These challenges further underscore the need for the affordable housing policies and programs that we outline in this book. To meet the needs of poor households, all levels of government will need to support the development of affordable housing. These initiatives may include the development of public or social housing, expansion of tax credit programs, additional use of project-based rental assistance, and a range of financial incentives to entice developers to construct more affordable housing.

Despite the numerous programs described in this book, a major issue for the affordable housing field is how to provide housing for the lowest income households with a single subsidy. Frequently, affordable housing projects require multiple subsidies and various sources of funding to be financially feasible. Coordinating these funding streams from different government agencies and private organizations is often complicated and time-consuming. Each funding source comes with its own set of restrictions and regulations, adding layers of bureaucracy to the process. This increased administrative burden increases the time to develop housing and raises development costs, as additional personnel and technical experts are needed to navigate these complexities. Relying on multiple subsidies and funding sources can also lead to uncertainty in financing affordable housing projects. Changes in funding priorities or availability can disrupt construction timelines and force developers to find alternative funding sources during construction, causing delays and increased costs. These challenges further complicate the already difficult task of creating affordable housing.

Note

1 Valuable LIHTC resources:

Scally, C.P., Gold, A., Hedman, C., Gerken, M, and DuBois, N. (2018). *The low-income housing tax credit: Past achievements, future challenges*. Urban Institute. Retrieved from www.urban.org/research/publication/low-income-housing-tax-credit-past-achievements-future-challenges

Scally, C.P., Gold, A., and DuBois, N. (2018). *The low-income housing tax credit: How it works and who it serves*. Urban Institute. Retrieved from www.urban.org/sites/default/files/publication/98758/lithc_how_it_works_and_who_it_serves_final_2.pdf

Kneebone, E. and Reid, C.K. (2021). *The complexity of financing low-income housing tax credit housing in the United States*. Terner Center for Housing Innovation, University of California Berkeley. Retrieved from https://ternercenter.berkeley.edu/wp-content/uploads/2021/04/LIHTC-Complexity-Final.pdf

References

Affordable Housing Finance. (2023). *The rise of state housing tax credits*. Retrieved from www.housingfinance.com/finance/the-rise-of-state-housing-credits_o

Biber, J. (2007). *Financing supportive housing with tax-exempt bonds and 4% low-income housing tax credits*. New York, NY: Corporation for Supportive Housing.

Brooks, M. E. (2007). *Housing trust fund progress report 2007*. Frazier Park, CA: Housing Trust Fund Project.

Center on Budget and Policy Priorities. (2021). *Policy basics: Public housing*. Retrieved from www.cbpp.org/research/policy-basics-introduction-to-public-housing

Center for Community Change. (2016). *The 2016 housing trust fund survey report*. Retrieved from www.housingtrustfundproject.org

Colburn, G., Acolin, A., & Walter, R. (2024). Subsidy overlap in federal housing policy. *Housing Policy Debate*.

Congressional Research Service. (2022). *Block grants: Perspectives and controversies*. Washington, DC: Congressional Research Service

Congressional Research Service. (2023). *The section 8 project-based rental assistance program*. Washington, DC: Congressional Research Service.

Connerly, C. E. (1990). Housing trust funds: New resources for low-income housing. *Journal of Housing*, 47(2), 96–104.

Couch, L. (2019). *Section 202: Supportive housing for the elderly*. Washington, DC: National Low-Income Housing Coalition.

Flink, C., Walter, R. J., & Xu, X. (2021). Policy diffusion in a redistributive policy: Affordable housing and state housing trust funds. *State and Local Government Review*, 53(3), 187–209.

Goetz, E. (2011). Where have all the towers gone? The dismantling of public housing in U.S. cities. *Journal of Urban Affairs*, 33(3), 267–287.

Goetz, E. (2013). The audacity of HOPE VI: Discourse and the dismantling of public housing. *Cities*, 35, 342–348.

Gramlich, E. (2023) *Rental assistance demonstration. 2023 advocates guide*. Washington, DC: National Low Income Housing Coalition.

Gress, T., Cho, S., & Joseph, M. (2016). *HOPE VI data compilation and analysis*. Washington, DC: Office of Policy Development and Research, U.S. Department of Housing and Urban Development.

Joice, P. (2017). HOPE and choice for HUD-assisted households. *Cityscape, 19*(3), 449–473.

Keightly, M. (2017). *An introduction to the low-income housing tax credit*. Washington, DC: Congressional Research Service.

Larsen, K. (2004). Florida's housing trust fund: Addressing the state's affordable housing needs. *Journal of Land Use & Environmental Law, 19*(2), 525–535.

Mueller, E. J., & Schwartz, A. (2008). Reversing the tide: Will state and local governments house the poor as federal direct subsidies decline? *Journal of the American Planning Association, 74*(1), 122–135.

National League of Cities. (2022). *Choice neighborhoods: HUD grants for people and places*. Retrieved from www.nlc.org/article/2022/09/02/choice-neighborhoods-hud-grants-for-people-places/#:~:text=In%20the%20past%20decade%2C%20Choice,housing%20units%20in%2040%20cities

National Low Income Housing Coalition. (2020). *HTF: The housing trust fund*. Retrieved from https://nlihc.org/sites/default/files/HTF_Factsheet.pdf

National Low Income Housing Coalition. (2024). *National housing trust fund*. Retrieved from https://nlihc.org/explore-issues/projects-campaigns/national-housing-trust-fund

Novogradac. (2024). *LIHTC equity pricing trends*. Retrieved from www.novoco.com/resource-centers/affordable-housing-tax-credits/lihtc-equity-pricing-trends

Scally, C., Gold, A., & DuBois, N. (2018). *The low-income housing tax credit: How it works and who it serves*. Washington, DC: Urban Institute.

Treskon, M., Reed, M., Popkin, S. J., & Champion, E. (2022). *Evaluation of the Rental Assistance Demonstration (RAD). Early findings on choice mobility implementation*. Washington, DC: Urban Institute.

Turner, M. A., & Kingsley, G. T. (2008). *Federal programs for addressing low-income housing needs*. Washington, DC: Urban Institute.

U.S. Department of Housing and Urban Development (HUD). (2010). *About HOPE VI*. Retrieved from www.hud.gov/program_offices/public_indian_housing/programs/ph/hope6/about#6

U.S. Department of Housing and Urban Development (HUD). (2024). *Project-based rental assistance. Summary of resources*. Washington, DC: Office of Housing, U.S. Department of Housing and Urban Development.

Chapter 9

Demand Side Housing Assistance

Overview of Demand Side Housing Assistance

In the last chapter, we outlined supply side approaches to creating or sustaining affordable housing. In Chapter 4, we outlined the historical trajectory of housing policy in the United States, which included a watershed transition in the 1970s and 1980s: rather than continuing to provide supply side assistance – in the form of public housing – the federal government began to provide purchasing power (in the form of housing vouchers) to eligible households that they could use to procure housing in the private market. This transition had a profound impact on policymaking in the United States and on the households that receive support. Under the new approach, the federal government no longer served as developer, operator, and landlord. Rather, the government assumed the role of funder in which it provides purchasing power to (or on behalf of) eligible households who receive housing assistance from the federal government. The supply side approach which dominated housing policy until the 1970s created actual housing units (in the form of large public housing projects) so these housing benefits were tied to specific units. Under the new demand side approach, housing assistance is frequently portable and tied to the recipient household.

A fundamental aspect of the demand side approach to housing support is that these policies and programs engage private owners and landlords and remove responsibility of property ownership and management from government agencies. Demand side rental assistance seeks to use government subsidy to close the housing affordability gap for eligible households. Support is delivered in the form of unrestricted cash assistance or a voucher that is tied to a specific purpose, in this case housing. In voucher programs, households contribute to the cost of their housing based on their financial capacity and the rental assistance (voucher) pays the difference between the household contribution and the market rent. The generosity of the support varies depending on the program/agency offering support, the level of household income, and local market rents.

The wisdom of this policy transition has been hotly debated and contested. Some argue that demand-side assistance is a more efficient and effective way to provide affordable housing than was public housing or rent control (Deng, 2010; Priemus et al., 2005). Waning public support for public housing also provided political momentum for the shift. Negative perceptions of public housing were shaped by concerns over the cost of production and management of government-owned housing which placed significant burdens on the public coffers. High profile "failures" such as Cabrini-Green in Chicago and Pruitt-Igoe in St. Louis led to increased stigmatization of public housing. These projects were neglected and became known as a symbol of failed government-owned housing. The beginning of the end for public housing was the Nixon Administration's moratorium on the construction of publicly-owned housing in 1973. But, in reality, the ideological push for more private market involvement began during the Johnson and Kennedy administrations as they experimented with promoting private investment in public housing. A thread of housing scholarship has countered the "failure" narrative of public housing. It argues that the vast majority of public housing operated well and served an important purpose, despite public perception that was shaped by a few, notorious public housing projects (Goetz, 2013).

The move toward demand-side supports accelerated in the early 1970s, at the same time that the Nixon moratorium was announced. The U.S. Department of Housing and Urban Development (HUD) launched three housing allowance experiments with 30,000 low-income households known as the Experimental Housing Allowance Program (EHAP). These experiments assessed the impact of allowances on the housing market, how households used the allowances, and program administration. A key recommendation from the EHAP demonstration was to phase out supply side subsidies and establish housing allowances, or demand side housing strategies, as the primary form of housing support to low-income households (Priemus et al., 2005). The Housing Choice Voucher program has become the primary source of housing support in the nation, and it can trace its lineage to the EHAP program of the early 1970s.

In this chapter, we break down demand-side supports into two categories: the Housing Choice Voucher Program and other federal and state supports and cash transfers. The HCV program is explicitly a housing program while other cash transfer programs are not earmarked exclusively for housing, nor are they managed by HUD. But, because unrestricted cash is fungible and can be used for housing expenses (the largest expenditure in most households' budgets), a discussion of these programs is essential when discussing policies designed to provide affordable housing.

Housing Choice Voucher Program

The largest demand side housing assistance program in the nation is the Housing Choice Voucher (HCV) program, which was established in 1974. The HCV program is funded by the U.S. Department of Housing and Urban Development (HUD) and is administered by local public housing authorities. The federal government provides over $20 billion in housing assistance through the HCV program that supports over 2.2 million households with the average subsidy around $950 per month (HUD, 2023b). It is the primary mechanism for providing deep subsidies – subsidies that house the lowest income households with high risk of housing insecurity. Local public housing authorities have some flexibility to establish preferences for vouchers based on local needs and context, such as households with children, older adults, or veterans. Public housing authorities can also convert up to 20% of their voucher capacity to create project-based vouchers (and up to 30% in special circumstances with approval). Most HCVs are tenant-based, which means that the voucher is attached to the household – if the household moves, the voucher goes with them. Project-based vouchers are tied to specific housing units and the current occupant (who must be eligible for a voucher) receives the benefit. A housing authority may use project-based vouchers to help households secure housing in areas where it is difficult to use tenant-based vouchers, such as high-cost cities where there are not abundant affordable housing options. Project-based vouchers may also be used to connect tenants with certain needs to properties that offer services such as physical and behavioral health or intensive care management.

The HCV program currently serves many older adults and people with disabilities on fixed incomes who do not participate in the labor force. For many voucher recipients, social security is their primary source of income. Figure 9.1 below highlights the demographics of HCV recipients. Households that use vouchers are frequently single older adults, people with disabilities, or female heads of household with children.

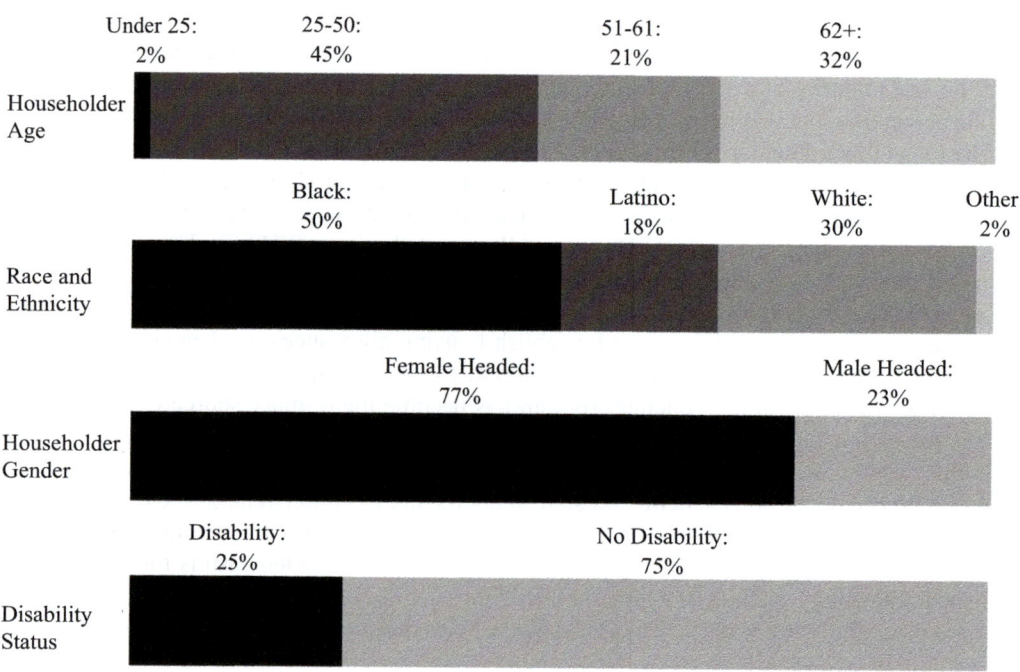

Figure 9.1 Characteristics of HCV Recipients
(Source: 2022 HUD's Picture of Subsidized Households)

The process of applying for and receiving a voucher is somewhat complicated. First, a household must add their name to a waitlist which is managed and maintained by the local housing authority. To be added to the waitlist, a household must be federally eligible based on household income. According to program rules, 75% of new HCV recipients must have extremely low incomes, which is defined as households that earn less than 30% of the area median income (AMI) (Dawkins & Jeon, 2017). The remaining 25% of vouchers may be distributed to households with incomes up to 80% of the AMI. But applying for an HCV does not, in most circumstances, yield immediate access to a voucher. Instead, the more common experience is spending years on the voucher waitlist. The long waits are a function of the inadequacy of the program at the federal level relative to the demonstrable need. Given its limited size, only one in four households that are eligible for federal housing support actually receive it (Poethig, 2014). Therefore, the demand for vouchers is far greater than their supply, hence, the multi-year waitlists. Given these dynamics, being selected to receive a voucher is like winning the lottery. But unlike a lottery, putting the voucher to productive use can be more difficult than simply cashing in one's lottery ticket.

Once a housing authority issues a voucher to a recipient, households begin the process of selecting a housing unit in the private market. The logic of the program is to provide flexibility to recipient households to allow them to find housing that meets their specific geographic and household preferences within the constraints of program guidelines. To be eligible to lease a unit to a voucher holder, landlords must pass an inspection conducted by the public housing authority that issued the voucher to confirm compliance with HUD's Housing Quality Standards. These standards are in place to ensure decent, safe, and sanitary housing conditions. If the unit inspection fails, the landlord is responsible for remediating all issues and can schedule a second inspection once the repairs have been completed. If the unit inspection fails the second time, the voucher recipient is advised to find a different unit. Landlords must comply with all provisions in the lease agreement and abide by local landlord-tenant laws.

While the intent of the program is to provide access to private market rental housing, voucher holders are constrained by the generosity of the program. A voucher can only be used to pay rents equal to HUD's fair market rents, which represent the 40th percentile of rent plus utilities in a metropolitan area (up to 50th percentile in some high-cost markets). These payment standards are based on a two-bedroom rental unit and adjusted for larger and smaller units. The local public housing authority sets payment standards that establish the maximum rent that a voucher holder can lease a unit for – typically between 90% to 110% of the fair market rent (HUD, 2023a). The purpose of these restrictions is to provide a standard level of housing quality for voucher recipients without subsidizing luxury housing. Therefore, the universe of potential housing units in which a voucher can be used are those units that meet the HUD criteria for housing quality that are priced below the prevailing HUD fair market rent and are unoccupied. A consequence of this constraint is that there are many high-priced neighborhoods that are not accessible to voucher recipients because the rental housing in those locations is more expensive than the voucher payment standards.

The housing search for voucher recipients can be complicated. While time periods vary by housing authority, households generally have up to 60 days to use their voucher to lease a housing unit, but in certain hard to rent cities, the period can be extended to 120 days or more. For voucher recipients, the housing search process concludes in one of two ways. For a fortunate set of voucher recipients, they are able to find a unit to lease using their voucher. A benefit of the HCV program is that a household can use the voucher in perpetuity to remain housed in that particular unit (or a subsequent unit) if they follow program guidelines (e.g., no lease violations, income does not exceed the maximum allowed). Successful lease-up, while far from perfect, is associated with a range of positive stable housing outcomes for those households (Gubits et al., 2016; Jacob & Ludwig, 2012; Mills et al., 2006). Unfortunately, for a significant percentage of voucher recipients, the housing search ends without successful lease-up of a housing unit. The most recent research suggests that only 60% of recipients are able to use their voucher to lease a housing unit (Ellen et al., 2024). For people who have waited for, in some cases, years to receive a voucher, the prospect of returning the unused voucher is an understandably frustrating outcome. Research highlights a range of reasons for difficult lease-up including, tight housing market conditions such as low vacancy rates (Ellen et al., 2024; Finkel & Buron, 2001), unwillingness of landlords to participate in the program (Garboden et al., 2018), and discrimination (Ellen et al., 2024; Tighe et al., 2017). A landlord must be willing to accept a voucher, pass a housing inspection, and charge a reasonable rent, so these extra administrative burdens make some landlords reluctant to participate in the program. The stigma associated with the program is also a disincentive for some landlords. Such stigmas have led to discriminatory actions by landlords based on source of income, where landlords are reluctant to rent to voucher holders especially in higher income neighborhoods. Many jurisdictions have passed source of discrimination laws to prevent discrimination against voucher recipients (Tighe et al., 2017), but enforcement and efficacy of such codes is limited (Freeman & Li, 2014).

Analyses of the voucher program yield mixed results. In some respects, the program has improved the experiences and outcomes of recipients, while there are also notable shortcomings. For those fortunate to receive a voucher and successfully lease a housing unit, the program has been successful in providing stable housing for millions of low-income households. Given the purchasing power boost of a housing voucher, the program reduces rent burdens of recipient households which then minimizes the risk of housing insecurity and homelessness, and reduces crowded housing situations (Gubits et al., 2016; Jacob & Ludwig, 2012; Mills et al., 2006). Beyond housing stability, the voucher program has been evaluated on its ability to allow households to move to neighborhoods that offer opportunity and promote economic mobility. On the positive side, voucher recipients report better housing quality and neighborhood outcomes than their unassisted peers – other low-income households that do not receive a voucher (Colburn, 2019; Pendall, 2000). Such a result should not be a surprise given the substantial purchasing power boost provided by the voucher. Furthermore, voucher recipients live in lower poverty neighborhoods compared to other HUD-assisted households such as those living in public housing (Hartung & Henig, 1997; Newman & Schnare, 2017; Pendall, 2000). Unfortunately, even though voucher recipients outperform their assisted and unassisted peers in terms of locational outcomes, they still live in relatively high poverty neighborhoods and in school districts with lower quality schools (Ellen et al., 2014). As noted earlier, access to high rent areas is limited given that only about 15% of rental units in high rent neighborhoods comply with the rent regulations of the voucher program (Collinson & Ganong, 2014).

On the negative side, the most glaring problem with the voucher program is its inadequacy. Because three out of four eligible households do not receive a voucher, there is substantial unmet need. But even among voucher recipients, many households struggle to find neighborhoods that provide accessible transit, quality schools, employment options, and other desirable amenities (Pendall, 2000; Walter & Wang, 2016). The spatial distribution of rental units that can be leased with a voucher is typically contained to racially segregated neighborhoods with high poverty that lack amenities, are unsafe, and have underperforming schools (Walter & Wang, 2016). There are a range of other critiques of the program. First, the program can be difficult for older adults or people with disabilities who have a very hard time navigating the search process for suitable housing in the private market (Pashup et al., 2005; Phinney et al., 2007). Second, some have argued that issuing vouchers creates inflationary pressure on rental housing, but the evidence of such a claim is mixed (Eriksen & Ross, 2015; Susin, 2002). Third, because many voucher recipients continue to reside in lower income and highly segregated neighborhoods, there is a fear that the program perpetuates a cycle of poverty by disincentivizing work because of the programmatic income and rental payment requirements (Riccio et al., 2017).

Because choice and locational preference is a fundamental attribute of the voucher program, a significant volume of research has been conducted on locational outcomes of voucher recipients. This body of research can be traced back to Chicago's Gautreaux Project in 1976. Gautreaux was a desegregation project created in response to litigation against the Chicago Housing Authority for racial discrimination in the siting of public housing in minority neighborhoods with high concentrations of poverty. The resolution of the litigation was a consent decree that required the Chicago Housing Authority to provide public housing residents with vouchers to relocate to other neighborhoods, including majority white neighborhoods.

The programmatic logic of Chicago's Gautreaux Project served as the impetus for the Moving to Opportunity (MTO) research demonstration by HUD. From 1994 to 1998, households who were living in public housing in five cities (Baltimore, Boston, Chicago, Los Angeles, and New York City) were randomly assigned to one of three groups: an experimental group that had to use a voucher in a neighborhood where the poverty rate was less than 10%; a group that could use a voucher to move to any neighborhood; or the control group that stayed in public housing. The purpose was to see how moving from areas of concentrated poverty improved economic conditions and health. Research over two decades later has shown that while MTO helped families move to areas with less poverty, there were few positive outcomes in terms of economic stability and health for parents, although recipients did report feeling safer in lower poverty neighborhoods (Ludwig et al., 2013). For children, the evidence was mixed. Early evidence suggested limited, immediate effects for children, but more recent research suggests that the long-run effects were more positive. As adults, children of parents who received a voucher and moved to lower poverty neighborhoods were more likely to attend college, less likely to be single parents, have higher incomes as adults, and live in lower poverty neighborhoods (Chetty et al., 2015). Furthermore, a recent experiment, Creating Moves to Opportunity, demonstrates these outcomes can be improved through customized housing search support and neighborhood information, relocation financial assistance, and connections to landlords (Bergman et al., 2024).

Other Rental Assistance Programs

At the time this book was being written, HUD was considering a new approach to providing rental assistance. Given the well-documented challenges associated with the HCV program, HUD is pursuing a direct rental assistance program that would provide cash, in the form of a prepaid debit card, directly to recipients to use for housing expenses. This approach differs meaningfully from the voucher program where the cash assistance is provided directly to the landlord. When payment is given directly to the landlord instead of the recipient, there is far more administrative burden for the landlord, and this has been demonstrated to deter participation in the voucher program. By providing cash directly to renter households, the hope is that such a program might make it easier for households to use this rental benefit to more easily access the housing that meets their needs and preferences. The current pilot of this program is being tested in Philadelphia and HUD is seeking other partners for the demonstration. If successful, HUD may petition Congress to fund a more extensive cash rental assistance demonstration (Cohen, 2023).

Besides federal rental housing assistance, there are also a range of programs managed by state and local governments. Because three out of four eligible households do not receive support from the federal government, households with significant unmet needs look to local sources for assistance. Both state and municipal governments offer rental and cash assistance programs to help recipient households afford quality, stable housing. Similar to the federal programs described above, state and local government programs provide direct housing assistance through rental subsidies or cash assistance that can be used for basic necessities such as shelter. The design, administration, and eligibility requirements vary by program and examples of these different structures are numerous. For instance, the City of Chicago has two programs; one provides relief in the form of rent payments to households currently in eviction court while the other program serves households at risk of homelessness. The scale and generosity of these programs tends to be far more limited than the programs managed by HUD.

State and local programs are not easy to track because their existence is volatile due to the fact that over 75% of the programs do not have long-term, dedicated funding sources (Aurand & Abdelhadi, 2023). Often states and cities use general revenue or special one-time funding allocations to fund rental assistance. Dedicated funding sources from real estate transfer taxes, sales tax, or interest on government accounts provide the most sustainable funding streams to support these programs. The National Low Income Housing Coalition created a database to catalogue state and city funded rental housing programs as a public service. Although the number of households served by these programs is unknown, 133 active rental assistance programs were identified in 38 states and 17 cities. Many of these programs assist persons experiencing or are at risk of homelessness. Other priority areas include persons with disabilities, persons with mental illness, extremely low-income households (income less than 30% AMI), youth, and victims of domestic violence (Aurand & Abdelhadi, 2023).

Income Transfers

The housing assistance programs described above provide support to low-income households that struggle to afford housing. There is another mechanism to provide support: unrestricted cash assistance. The federal government operates a number of cash assistance programs for households in a range of different situations. These programs provide cash that can be used for a range of household expenditures, including housing. Therefore, elected leaders have a range of policy levers to pull if they desire to provide for greater affordability for needy households. In this section, we cover some of the more significant cash assistance programs, including the Earned Income Tax Credit (EITC), Temporary Assistance for Needy Families (TANF), the Supplemental Security Income (SSI), and Social Security Disability Insurance (SSDI).

Temporary Assistance for Needy Families (TANF)

TANF is a federal block grant program administered by states to aid households that struggle to meet basic needs. When people refer to "welfare" in general terms, they are typically referring to this program. In 1996, TANF replaced the original cash assistance welfare program, Aid to Families with Dependent Children, that had been in place since 1935. TANF came into existence as part of President Bill Clinton's highly publicized welfare reform initiative and became law with the passage of the Personal Responsibility and Work Opportunity Reconciliation Act of 1996. The stated goal of welfare reform was to reduce government assistance by incentivizing more people to enter the labor force. Consistent with this policy objective, work requirements are a hallmark of TANF. Another key feature of welfare reform was significant delegation of responsibility to individual states. Rather than the federal government mandating certain benefit levels for the nation as a whole, TANF is administered at the state level. As a result, eligibility rules and program generosity vary from state to state. Some states are more generous with their benefits than others. In September of 2022, there were about 2.8 million TANF recipients, of which two million were children (CRS, 2023).

As a cash assistance program, TANF has two primary drawbacks. First, one of the primary purposes of cash assistance is to help families afford housing and other basic necessities, but the assistance is frequently insufficient to cover the full cost of rent, much less additional household expenditures. The median monthly TANF benefit in 2022 was $492 (CBPP, 2023), which covers only a small proportion of essential household expenses. Second, another attribute of TANF is that it is time limited; recipient households are not allowed to receive benefits for their entire lives. Households are limited to five years of participation in the program. Recipients must also participate in work activities while receiving benefits, which can be a challenge for single parent households that also have childcare responsibilities. Therefore, the program is limited in its ability to provide long-term financial support.

Earned Income Tax Credit (EITC)

The EITC provides a tax break to low-income households to offset taxes that are owed. To take advantage of the EITC, a taxpayer must have earned income during the tax year, but total earnings cannot be above a specified amount for the year. As an example, to qualify for EITC in 2022, the earned maximum adjusted gross income was $59,187 with three dependents and filing jointly, and the investment income maximum was $10,300. There are other stipulations, such as the taxpayer must be a U.S. citizen or a foreign-born, non-U.S. citizen that has lived in the United States for at least six months out of the year, and be 19 years of age or older. The amount of the credit which reduces income taxes owed varies by filing status, number of children or dependents, whether the taxpayer or any of their dependents have a disability, and the annual earned income for the tax year. In 2022, the credit range was from several hundred dollars up to almost $7,000. If the credit (or benefit) exceeds the tax liability, the household receives a cash refund from the Internal Revenue Service (IRS). Therefore, the EITC not only decreases the tax burden of low-income households, but it can also be a source of additional cash assistance. As of December 2022, 31 million workers received roughly $64 billion in EITC, with an average amount of $2,043 per worker (IRS, 2023). Many poverty scholars suggest that the EITC, and a potential expansion of EITC, is one of the country's best poverty alleviation policies for families with children (Acs & Werner, 2021; CBPP, 2023; Hendricks & Roque, 2021).

Social Security Disability Insurance (SSDI) and Supplemental Security Income (SSI)

SSDI is a direct cash transfer administered by the Social Security Administration. The beneficiaries of this program are individuals that have a qualifying medical condition and have been previously employed in a job which pays into the social security system. Individuals that have been unable to work for more than a year due to the medical condition receive monthly payments. Surviving spouses or unmarried children may be eligible for benefits after the primary recipient is deceased. The monthly benefit varies and depends on the individual's work history and earnings. There are more than 8 million recipients of SSDI and approximately 18% of recipients are surviving spouses or children (CBPP, 2021a).

SSI provides a direct cash transfer to low-income older adults and people with disabilities and is administered by the Social Security Administration. It was created to assist older adults, the blind, and disabled individuals that do not qualify for SSDI. To be eligible for SSI, recipients must have limited income and resources. SSI federal benefits are the same nationwide: a monthly benefit of $914 for one person and $1,371 for couples in 2023. However, the amount may vary based on where you live and how much income you earn since some states offer a supplement. In 2023, there were more than 7.4 million recipients of SSI as reported by the Social Security Administration. Since SSI is a cash transfer, many households use this benefit to help cover monthly household expenditures, including housing costs. Relative to TANF, SSI is more generous and serves two and half times as many recipients. SSI can be used with SSDI. When individuals are eligible for both, this is referred to as concurrent benefits. In 2020, there were over 1 million households receiving concurrent benefits (SSA, 2020). Many households use SSI and SSDI to help cover monthly housing costs.

State and Local Income Transfer Programs

Another form of cash assistance is a basic income program where recipients receive unrestricted cash assistance. Over the last ten years, there has been a proliferation of basic income experiments throughout the United States, and internationally. The logic behind these programs is to provide unrestricted cash to households in need. These programs have been shown to be an effective emergency response for households and individuals experiencing housing insecurity and homelessness. One of the most important findings from these studies is that recipient households use the cash assistance on basic household necessities, such as housing and food, instead of temptation goods such as alcohol and drugs (Dwyer et al., 2023; Doussard, 2023).

In 2019, the City of Stockton in California piloted a no strings attached program which provided 100 residents $500 a month. Since that time, over 40 programs in California have either been launched or are in the process of being initiated. Breathe, Los Angeles County's Guaranteed Income Program, provides households living in poverty $1,000 per month for three years and 200 youth aging out of foster care with $1,000 for two years. Since many of these programs were recently implemented, program evaluation is underway. Short-term findings reveal positive outcomes. Denver's basic income project, initiated in 2021 and focused on unhoused individuals, has produced positive housing and wellbeing outcomes for program participants (The Center for Housing and Homeless Research, 2023).

How recipients use cash assistance is an important contribution to the study of affordable housing policy. The largest housing assistance program in the United States, the Housing Choice Voucher program, was developed with a healthy dose of paternalism. One of the reasons for issuing vouchers rather than cash is that a voucher constrains its use for only a designated purpose. Implicit in that policy design is a concern that, left to their own devices, recipient households may use cash assistance in other, unproductive ways. A consequence of this paternalism is the administrative infrastructure and burden associated with the program. The results of these basic income studies – highlighting that low-income households use cash assistance in productive ways – opens the conversation to wider use of cash-based rental assistance which could have significant implications for program design and administrative oversight. Doussard (2023) also argues that the proliferation of basic income experiments has created an opening for greater emphasis on cash assistance programs for low-income households.

Vignette

The following vignette is based on actual experiences with public assistance. It highlights how challenging it can be for a household even when it receives multiple sources of income and housing support.

The Johnson family lives in a mid-sized city in the western United States. Denise Johnson is a single mother with five children. Her husband – who was the primary wage earner in the household – recently passed away. Unfortunately, Denise has no familial support, so she is left to care for her five children on her own. Two of the children are in elementary school while the other three are younger than five years old. Denise is not employed because the costs of childcare would exceed the income that she could earn from employment. Denise's only source of income is TANF, which provides about $7,000 per year (approximately $585 per month). For a household of this size, this level of income represents about 10% of the Area Median Income. Given that her rent is about $1,000 per month, she does not have sufficient income to afford housing, much less the other household expenses. Denise and her children have been forced to move frequently in and out of homeless shelters and temporary housing situations following the death of her husband.

Eventually, Denise was fortunate enough to receive a Housing Choice Voucher from the local housing authority. Denise is one of the lucky ones – three out of four eligible households don't receive a voucher. She was able to use the voucher to lease a rental unit in her community and the housing authority pays approximately 90% of the rent and Denise pays the remaining $100 per month. While the voucher is incredibly valuable, Denise still faces *the reality of having to pay for all of her other household expenditures with the $5,800 of funds from TANF after making her contribution toward rent. Obviously, this amount is insufficient to cover even the most basic needs for a family of six. Other programs, such as the EITC may provide Denise with additional funds, but the financial picture for her family remains bleak.*

This vignette highlights the challenge for low-income households, especially those that face an adverse event such as the death of a spouse. What is also important to underscore is that – from the perspective of the public assistance system in the United States – the Johnsons are one of the lucky households. There are millions of low-income households that do not receive the financial support and stability provided by a housing voucher. Financial hardship and homelessness are unfortunate, but inevitable, outcomes of this system.

Conclusion

We conclude by assessing the structure and adequacy of demand side housing assistance. First, as highlighted above, demand side housing assistance can be structured with a stated use – the benefit can only be used to fund housing expenditures – such as the voucher program. Cash-based assistance, which is unrestricted, allows households to use the funds in any way that they choose to do so. These unrestricted funds can come from traditional welfare programs, such as TANF or SSI, but may also be delivered via basic income programs that could be funded by private or governmental sources. The recently proposed HUD demonstration of cash-based rental assistance marks a departure for the government in the provision of housing support. All prior programs either delivered a housing unit to households (either via public housing or LIHTC) or provided purchasing power to households in the form of a voucher that can only be used for housing. The many state and local rental assistance programs are further examples of cash-based housing support. One of the important areas of debate in housing policy will continue to be the structure of these programs: should we provide benefits with a stated use, or should we provide unrestricted cash assistance?

Second, we must assess the adequacy of these programs. There is overwhelming evidence to suggest that both the generosity and reach of these programs are inadequate. The Housing Choice Voucher program is, by far, the most generous program, requiring a household to pay approximately 30% of their income toward rent. The primary issue with the voucher program is its limited scope. Only one in four eligible households receive the benefit (Center on Budget and Policy Priorities, 2021). Therefore, three out of four poor and eligible households must find other sources of money to pay the rent. The other programs, such as TANF, SSI, and SSDI serve more households, but the level of assistance is inadequate to support all of the expenses of a household, especially in high-cost locations. Many households, therefore, must piece together different supports to cover basic necessities. In sum, the United States lacks a comprehensive housing policy that adequately supports low-income households, and the consequences are tragic, but predictable. Nearly a quarter of all rental households pay more than 50% of their monthly income for housing and, in any given year, over half a million people experience homelessness on a given night (Joint Center for Housing Studies, 2023).

Critics of a more robust program of housing assistance have highlighted a number of concerns. First, unrestricted cash assistance opens the door to recipients using public assistance to purchase non-essential goods and services. There are some who do not want public funds to be used to purchase alcohol or drugs, gamble, or to otherwise spend frivolously. As noted above, the recent evidence from the basic income experiments suggests that the cash assistance is used on households' necessities such as housing and food. Another argument against a more robust system of cash or rental assistance is the concern that such supports reduce the willingness to work. Such concerns are long-standing and are a key reason why work requirements were included in the structure of the TANF program in 1996. Again, the basic income experiments provide evidence that such concerns may be overblown. Multiple studies show that work effort did not decline after receipt of cash assistance (Bastagli et al., 2016; Gentilini et al., 2019; Hoynes & Rothstein, 2019; Marinescu, 2018). While it is unlikely that opposition to cash-based assistance will cease immediately, a growing evidence base about the efficacy and efficiency of cash assistance could provide momentum for further expansion of these programs in the future.

References

Acs, G., & Werner, K. (2021). *How a permanent expansion of the child tax credit could affect poverty*. Washington, DC: Urban Institute.

Aurand, A., & Abdelhadi, S. (2023). *State and local investments in rental housing: A summary of findings from the 2023 rental housing programs database*. Washington, DC: National Low Income Housing Coalition.

Bastagli, F., Hagen-Zanker, J., Harman, L., Barca, V., Sturge, G., & Schmidt, T. (2016). *Cash transfers: What does the evidence say?* Overseas Development Institute.

Bergman, P., Chetty, R., DeLuca, S., Hendren, N., Katz, L. F., & Palmer, C. (2024). Creating moves to opportunity: Experimental evidence on barriers to neighborhood choice. *American Economic Review*, *114*(5), 1281–1337.

Center on Budget and Policy Priorities (CBPP). (2021). *Chart book: Social security disability insurance*. Retrieved from www.cbpp.org/research/social-security/social-security-disability-insurance-0

Center on Budget and Policy Priorities (CBPP). (2021a). *Increases in TANF cash benefit levels are critical to help families meet rising costs*. Retrieved from www.cbpp.org/research/income-security/increases-in-tanf-cash-benefit-levels-are-critical-to-help-families-meet-0

Center on Budget and Policy Priorities (CBPP). (2021b). *More housing vouchers needed for people with behavioral health challenges*. Retrieved from www.cbpp.org/research/housing/more-housing-vouchers-needed-for-people-with-behavioral-health-challenges

Center on Budget and Policy Priorities (CBPP). (2023). *Policy basics: The earned income tax credit*. Retrieved from www.cbpp.org/research/policy-basics-the-earned-income-tax-credit

Chetty, R., Hendren, N., & Katz, L. F. (2015). *The effects of exposure to better neighborhoods on children: New evidence from the moving to opportunity experiment*. Retrieved from https://opportunityinsights.org/wp-content/uploads/2018/03/mto_paper.pdf

Cohen, R. M. (2023). *A bold new experiment in giving renters cash. Would money help tenants more than housing vouchers?* Retrieved from www.vox.com/2023/9/12/23864165/affordable-housing-voucher-program-hud-federal-government-section-8

Colburn, G. (2019). The effect of market conditions on the housing outcomes of subsidized households: The case of the US voucher programme. *Housing Studies*, *34*(9), 1465–1484.

Collinson, R. A., & Ganong, P. (2014). The incidence of housing voucher generosity. Working paper. Retrieved from https://scholar.harvard.edu/files/ganong/files/collinsonganong105.pdf

Congressional Research Service (CRS). (2023). *The temporary assistance for needy families (TANF) block grant: Responses to frequently asked questions*. Retrieved from https://sgp.fas.org/crs/misc/RL32760.pdf

Dawkins, C., & Jeon, J. (2017). *Rent burden in the Housing Choice Voucher program*. U.S. Department of Housing and Urban Development, Office of Policy Development and Research.

Deng, L. (2010). The cost-effectiveness of the low-income housing tax credit relative to vouchers: Evidence from six metropolitan areas. *Housing Policy Debate*, 16(3–4), 468–511.

Doussard, M. (2023). Viral cash: Basic income trials, policy mutation, and post-austerity politics in US cities. *Environment and Planning A: Economy and Space*.

Dwyer, R., Palepu, A., Williams, C., Daly-Grafstein, D., & Zhao, J. (2023). Unconditional cash transfers reduce homelessness. *Proceedings of the National Academy of Sciences, 120*(36), 1–9.

Ellen, I. G., O'Regan, K., & Strochak, S. (2024). Race, space, and take up: Explaining housing voucher lease-up rates. *Journal of Housing Economics*, 63.

Eriksen, M. D., & Ross, A. (2015). Housing vouchers the price of rental housing. *American Economic Journal: Economic Policy, 7*(3), 154–176.

Finkel, M., & Buron, L. (2001). *Study on section 8 voucher success rates: Volume I: Quantitative study of success rates in metropolitan areas.* Cambridge, MA: Abt Associates.

Freeman, L., & Li, Y. (2014). Do source of income anti-discrimination laws facilitate access to less disadvantaged neighborhoods? *Housing Studies, 29*(1), 88–107.

Garboden, P. M. E., Rosen, E., DeLuca, S., & Edin, K. (2018). Taking stock: What divides landlord participation in the Housing Choice Voucher program. *Housing Policy Debate, 28*, 6, 979–1003.

Gentilini, U., Grosh, M., Rigolini, J., & Yemtsov, R. (2019). *Exploring universal basic income: A guide to navigating concepts, evidence, and practices.* The World Bank.

Goetz, E. (2013). *New deal ruins: Race, economic justice, & public housing policy.* Ithaca, NY: Cornell University Press.

Gubits, D. et al. (2016). *Family options study: Three-year impacts of housing and services interventions for homeless families.* Washington, DC: U.S. Department of Housing and Urban Development.

Hartung, J. M., & Henig, J. (1997). Housing vouchers and certificates as a vehicle for deconcentrating the poor: Evidence from the Washington, D.C. metropolitan area. *Urban Affairs Review, 32*, 403–419.

Hendricks, G., & Roque, L. (2021). *An expanded child tax credit would lift millions of children out of poverty.* Washington, DC: Center for American Progress. Retrieved from www.americanprogress.org/issues/economy/reports/2021/02/23/495784/expanded-child-tax-credit- lift-millions-children-poverty/.

Hoynes, H., & Rothstein, J. (2019). *Universal basic income in the us and advanced countries.* National Bureau of Economic Research.

Internal Revenue Service (IRS). (2023). *EITC reports and statistics.* Retrieved from www.irs.gov/credits-deductions/individuals/earned-income-tax-credit/eitc-reports-and-statistics#:~:text=Nationwide%20as%20of%20December%202022,about%20%2464%20billion%20in%20EITC.

Jacob, B. A., & Ludwig, J. (2012). The effects of housing assistance on labor supply: Evidence from a voucher lottery. *American Economic Review, 102*(1), 272–304.

Joint Center for Housing Studies. (2023). *The state of the nation's housing.* Retrieved from www.jchs.harvard.edu/sites/default/files/reports/files/Harvard_JCHS_The_State_of_the_Nations_Housing_2023.pdf

Ludwig, J., Duncan, G. J., Gennetian, L. A., Katz, L. F., Kessler, R. C., Kling, J. R., & Sanbonmatsu, L. (2013). Long-term neighborhood effects on low-income families: Evidence from moving to opportunity. *American Economic Review, 103*(3), 226–231.

Marinescu, I. (2018). *No strings attached: The behavioral effects of U.S. unconditional cash transfer programs.* National Bureau of Economic Research.

Mills, G., Gubits, D., Orr, L., Long, D., Feins, J., Kaul, B., Wood, M., Jones, A., Cloudburst Consulting Associates, the QED Group. (2006). *The effects of housing vouchers on welfare families.* Washington, DC: U.S. Department of Housing and Urban Development, Office of Policy Development and Research.

Newman, S. J., & Schnare, A. B. (1997). '… And a suitable living environment': The failure of housing programs to deliver on neighborhood quality. *Housing Policy Debate, 8*, 703–741.

Pashup, J., Edin, K., Duncan, G. J., & Burke, K. (2005) Participation in a residential mobility program from the client's perspective: Findings from Gautreaux Two. *Housing Policy Debate, 16*(3–4), 361–392.

Pendall, R. (2000). Why voucher and certificate users live in distressed neighborhoods. *Housing Policy Debate, 11*(4), 881–910.

Phinney, R., Danziger, S., Pollack, H. A., & Seefeldt, K. (2007) Housing instability among current and former welfare recipients. *American Journal of Public Health, 97*(5), 832–837.

Poethig, E. C. (2014). *One in four: America's housing assistance lottery.* Washington, DC: Urban Institute.

Priemus, H., Kemp, P. A., & Varady D. P. (2005). Housing vouchers in the United States, Great Britain, and the Netherlands: Current issues and future perspectives. *Housing Policy Debate, 16*(3–4), 575–609.

Riccio, J. A., Deitch, V., & Verma, N. (2017). *Reducing work disincentives in the housing choice voucher program: Rent reform demonstration baseline report.* Washington, DC: U.S. Department of Housing and Urban development, Office of Policy Development and Research.

Social Security Administration (SSA). (2020). *Annual statistical report on the social security disability insurance program, 2020.* Retrieved from www.ssa.gov/policy/docs/statcomps/di_asr/2020/sect05.html

Susin, S. (2002). Rent vouchers and the price of low-income housing. *Journal of Public Economics, 83,* 109–152.

The Center for Housing for Homeless Research. (2023). *Denver basic income project. Quantitative income report.* Retrieved from https://static1.squarespace.com/static/64f507a995b636019ef8853a/t/651ef48d225414181ea2b907/1696527503685/DBIP+Quantitative+Report+One-Pager.pdf

Tighe, R.J, Hatch, M. E., & Mead, J. (2017). Source of income discrimination and fair housing. *Journal of Planning Literature, 32*(1), 3–15.

U.S. Department of Housing and Urban Development (HUD). (2023a). *The housing choice voucher program guidebook.* U.S. Department of Housing and Urban Redevelopment, Office of Public and Indian Housing.

U.S. Department of Housing and Urban Development (HUD). (2023b). *Housing choice voucher dashboard.* Retrieved from www.hud.gov/program_offices/public_indian_housing/programs/hcv/dashboard

Walter, R. J., & Wang, R. (2016). Searching for affordability and opportunity: A framework for the Housing Choice Voucher program. *Housing Policy Debate, 26*(4–5), 670–691.

Chapter 10

Affordable Homeownership

Homeownership in the United States

Homeownership is a key component of the American Dream and has long been a symbol of financial success in the United States. It offers housing security through stable mortgage payments and control over how long one chooses to live at the property. Homeownership serves as a vehicle for wealth accumulation through increased property values and a paydown of debt over time. As long as zoning laws, building codes, and any other rules and regulations that direct the use of the property are followed, homeowners are free to make improvements to their home and decisions about how the property is used.

Homeownership is the dominant form of tenure in the U.S. The homeownership rate has been above 60% since the 1960s. However, homeownership rates vary by race and income (see Figure 10.1). White households have the highest homeownership rates of all racial and ethnic groups; nearly three-quarters of all white households own their own home. Households of color have lower rates of homeownership and this has been consistent over time due to various forms of discrimination that have limited access to homeownership (Wachter & Acolin, 2022). The

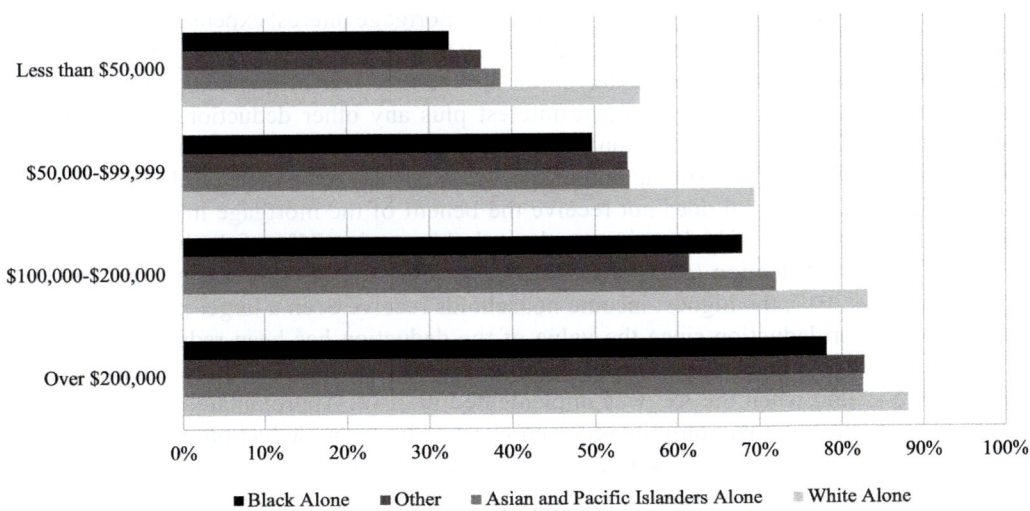

Figure 10.1 Homeownership Rates by Household Race and Income
(Source: American Housing Survey, U.S. Census Bureau)

homeownership rate discrepancy is most stark between Black and white households, especially those at lower income levels.

The U.S. housing system has been built to incentivize and promote property ownership. Home equity is the largest source of household wealth in the United States. For the average household, home equity comprises about half of a household's overall wealth (Board of Governors of the Federal Reserve System, 2023). Determined to protect this significant financial interest, homeowners frequently make decisions that they believe will protect their property value, such as support zoning regulations that restrict housing density in residential neighborhoods and similar policies that are associated with the "NIMBY" (Not In My Backyard) movement (Scally, 2012). Consequently, homeowners are more likely to be politically active and vote to protect their interests (McCabe, 2016). The vast majority of public officials are homeowners, while renters are significantly underrepresented in public office (Einstein et al., 2023). These tendencies result in overrepresentation of homeowner involvement in the political process, which makes it difficult to reform preferential treatment of homeownership. Thus, homeowners have been significant beneficiaries of our nation's housing policies. One of the most prominent examples is the mortgage interest deduction, which was accidentally included in the Revenue Act of 1913 but has become one of the most defended and expensive tax subsidies that disproportionately benefits middle- and higher-income households (Ventry, 2009).

The mortgage interest deduction allows homeowners to reduce their taxable income based on the amount of mortgage interest paid during the year. In 2022, households were allowed to deduct interest on mortgages up to a loan limit of $750,000. For example, suppose a household has a $400,000 mortgage with a loan term of 30 years and fixed interest rate of 5%. In the first year of this loan, $19,866 of the household's mortgage payment is to pay interest. If the household earned $150,000 in 2022 in wages, their taxable income may be reduced by the amount of interest paid depending on filing status, overall deductions, the state of residence, and several other factors. The deduction provides a considerable tax incentive for homeowners but not all homeowners enjoy the benefit. The deduction largely benefits homeowners with larger loans.

For households with smaller loans, the amount of mortgage interest expense may not qualify for the deduction. Suppose a household has a $150,000 loan with the same terms as above – 30-year loan term with a 5% fixed interest rate. In the first year of this loan, the homeowner paid $7,450 in interest. If the mortgage interest plus any other deductions do not exceed $12,950 (the 2022 standard deduction for a single person which is granted to taxpayers to reduce their taxable income without listing specific expenses), then the household would take the standard deduction and does not receive the benefit of the mortgage interest deduction. Therefore, the top 20% of highest income households receive 75% of the benefits from the mortgage interest deduction (Congressional Budget Office, 2013). Following the Tax Cuts and Jobs Act of 2017, the highest income households receive even a larger proportion of the mortgage interest deduction since the value of the deduction has been reduced pursuant to the legislation. The bill nearly doubled the standard deduction from $6,350 for a single person in 2017 to $12,000 when the Act was implemented in 2018. Approximately 77% of households in the highest income quintile received benefits from the deduction in the form of an average 1.5% reduction of their income (before taxes), compared to 19% in the lowest income quintile who received benefits of 0.2% of their income (before taxes) (Congressional Budget Office, 2021).

Most households that use the mortgage interest deduction would likely own their homes irrespective of the deduction, and the deduction is poorly designed to expand homeownership

(IASP & NLIHC, 2017). The tax benefit has been an inadequate tool for breaking down barriers for low- or moderate-income renters to enter homeownership. There is evidence that the mortgage interest deduction works against its intended purpose of expanding homeownership and instead incentivizes larger home purchases for the wealthy and increases home prices (Hanson, 2012; Sommer & Sullivan, 2017). Not only does this contribute to continued unaffordability in the housing market, but it also provides a substantial housing subsidy to the wealthiest households at the expense of those most in need, further exacerbating disparities. The Center on Budget and Policy Priorities reports that the highest income households received four times the housing benefits of low-income households (Fischer & Sard, 2017). The foregone revenue that the government would collect were it not for the mortgage interest deduction is greater than the amount of money spent on the Housing Choice Voucher program, the largest federal low-income rental assistance housing program in the nation. As one of the costliest expenditures in the tax code, the forgone revenue from the deduction is larger than the entire Department of Housing and Urban Development's budget (McCabe, 2018).

White, higher-income households not only have the highest homeownership rates but also have captured a disproportionate share of benefits from owning a home. Since homeownership is the largest asset and the primary investment strategy for most Americans, the observed racial wealth gap can, in part, be attributed to differences in rates of homeownership. Figure 10.2 highlights the significant wealth gap between white households and households of color. The net worth of white households is more than five times that of Black and Hispanic households since 1989.

Households that have had limited access to homeownership due to socioeconomic status and racial discrimination, have not benefitted at the same rate while other households have accumulated trillions of dollars of wealth as home values have increased. Compounding the inequity,

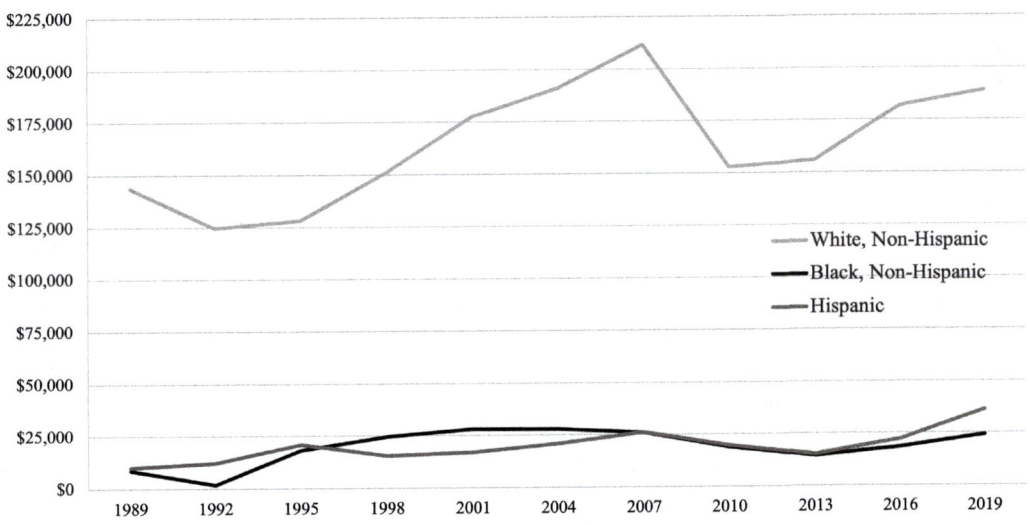

Figure 10.2 Household Net Worth by Race or Ethnicity

(Source: Survey of Consumer Finances, Federal Reserve Board) Note: Net worth refers to the difference between families' gross assets and their liabilities.

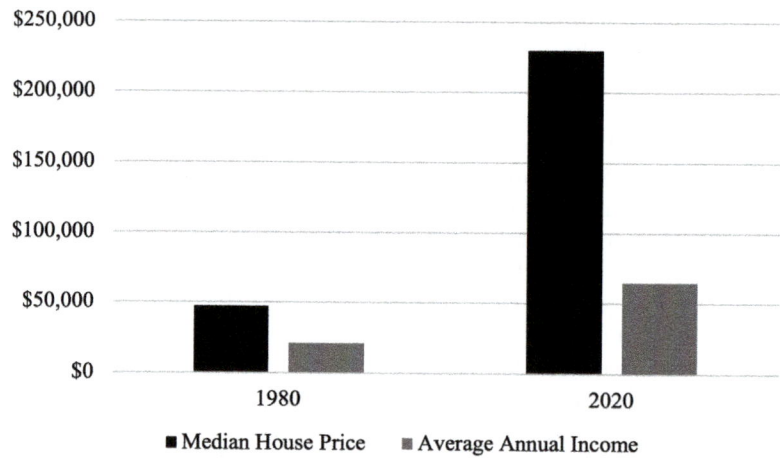

Figure 10.3 Home Values Outpace Income, 1980 v. 2020
(Source: U.S. Census Bureau)

accessing homeownership now is even more difficult given increasingly high home values and the highly unequal distribution of wealth in this country. Year-over-year home price appreciation continues to outpace wage growth, making it difficult for first-time homebuyers, and often impossible for moderate- and low-income households, to enter homeownership. In 1980, the median house price in the United States was only about double that of the average annual income (see Figure 10.3). In 2020, the median house price was three and a half times higher than the average annual income. The widening gap between income and home costs represents a major hurdle for households seeking to become homeowners.

First-time homebuyers and low-income households have been especially impacted by skyrocketing house prices. Higher prices increase down payment requirements and the amount of money needed for closing costs. The down payment, the most significant barrier to entering homeownership, has reached an all-time high. For homes priced under $350,000 in 2022, the average down payment was over $30,000 (Pradhan, 2022). Down payments in urban areas, where more than 80% of the population resides, are even more out of reach due to high home values and associated costs. Closing costs for buyers, which the homebuyer must pay on top of a down payment, are generally between 3% to 6% of the purchase price (Woodward, 2008). For a home priced around $350,000, closing costs alone are between $7,000 to $21,000. The down payment plus closing costs averages nearly $45,000 for a $350,000 home which is prohibitively expensive for many low- and middle-income homeowners. Access to homeownership is also restricted by income and credit (Acolin et al., 2016). Debt to income ratios prevent low-income households from entering homeownership. Low credit scores make mortgage loans more expensive or entirely unattainable (Dey & Brown, 2020).

The Survey of Consumer Finances indicates that the average savings for lower income and younger households in the U.S. is under $5,000. When compared to average closing costs of $45,000, this explains why homeownership is difficult to attain for millions of households (Perez, 2022). Beyond the cost to purchase a home, there are additional costs, including routine maintenance, major repairs, insurance, property taxes, landscaping and lawn care, and home improvements that increase the cost of homeownership. The remainder of this chapter provides an overview of the various tools that can provide access to homeownership for low-income households.

Homeownership Assistance Programs

There are federal, state, and local programs offered by governmental agencies, non-profits, and the private sector to make homeownership more affordable. These programs include down payment assistance, closing cost assistance, matched savings programs, mortgage financing, and tax credits.

Down Payment Assistance

When you buy a home, a certain percentage of the purchase price is required at the time of closing to take possession of the property, known as a down payment. The standard is 20% down because this amount proves to lenders that the homeowner has a vested financial interest in the property. A 20% down payment mitigates risk for the lender so they can offer more favorable financing terms for the mortgage. To purchase a home for $350,000 with a 20% down payment will require $70,000 of cash plus additional closing costs. As mentioned earlier in the chapter, most Americans, especially lower income and younger households, do not have this level of savings.

Over time, the minimum down payment required to purchase a home has declined in recognition of the sizable amount of upfront cash that is needed. Many conventional loans, which are loans that aren't guaranteed or insured by a government agency, only require 3 to 5% down without any additional penalties or fees. According to the National Association of Realtors, in 2021 the typical down payment for first-time buyers was 7% of the purchase price (Lautz, 2022). Beyond conventional loans, there are also mortgages provided by government entities, such as the Veteran's Administration and U.S. Department of Agriculture, that require no down payment.

In recognition of the substantial cash needs, down payment assistance programs provide the buyer with funds to help cover the cost of the down payment. This type of assistance may consist of a grant or loan from either government entities or non-profits, and the amounts vary greatly depending on the program. Some down payment assistance is structured as a grant and does not need to be repaid. However, it is more common for down payment assistance to be provided as a loan. There are many types of loans including forgivable loans (which don't need to be paid back), deferred-payment loans, and low-interest loans.

The Downpayment Toward Equity Act, a proposed federal down payment assistance program, was introduced in 2021 and is being debated in the U.S. Congress. If passed into law, the Act would provide first-generation, first-time moderate- and low-income homebuyers up to $25,000 as a cash grant to use at closing. The program would require recipients to live in their homes as their primary residence for at least five years to be exempt from repayment. Prior to the passage of this bill, buyers who need assistance now may consider thousands of other programs that exist through state and local governments, nonprofits, charities, employers, and the private sector that offer down payment assistance. Despite the prevalence of downpayment assistance programs, the availability of support is inadequate to the demand for such programs.

Closing Cost Assistance

Closing costs are the fees related to the purchase of a home beyond the down payment. These costs include mortgage origination, title work, inspection, appraisal, and escrow agent and attorney fees. Closing costs have increased over the years because many of the fees are directly tied to the home's purchase price. For a buyer, these costs range from 2% to 6% of the purchase price. Revisiting the example above, for a home purchased for $350,000, closing costs may be as high as $21,000. Therefore, down payment assistance programs often provide funds to support both closing costs and the down payment, while there are programs specifically designed to cover only closing costs, which are structured as grants or loans similar to down payment assistance programs. For example, the Downpayment Toward Equity Act includes a provision that allows up to $3,000 of the $25,000 grant to be used specifically for closing costs.

Matched Savings Programs

Matched savings programs, also known as individual development accounts, provide a financial incentive to households to save for specific goals that are tied to large expenses such as post-secondary education, starting a business, purchasing a car, or buying a home. The typical match rates range from 1:1 to 2:1. For example, for programs that provide a 2:1 match, if a participant saves $2,000 in a specified period, they receive $4,000 from the sponsoring entity for a total savings of $6,000 at the end of that period. Like programs that offer down payment and closing cost assistance, matched savings programs have criteria for eligibility and program guidelines that vary by program. For buying a home, matched savings funds are typically applied toward the down payment and closing costs. There are hundreds of entities that offer matched savings programs in the U.S. including credit unions, state and local governmental agencies, and national and local non-profit organizations.

The logic for matched savings programs in the United States originated in the early 1990s from Michael Sherraden's book, *Assets and the Poor New American Welfare Policy*. As part of the most significant welfare reform act – Personal Responsibility and Work Opportunity Reconciliation Act of 1996 – the federal Assets for Independence (AFI) program was authorized in 1998 and is administered by the U.S. Department of Health and Human Services to support the implementation of individual development account programs across the nation. Although the AFI program was last funded in 2016 and aided less than 100,000 households, it was the impetus for the creation of hundreds of matched savings programs. Despite the fact that matched savings programs have been beneficial for numerous individual households, research has indicated that their impact may be limited for long-term outcomes such as overall wealth generation and reduction in poverty rates (Mills et al., 2008).

Mortgage Financing

A home mortgage can be structured in ways to help lower income households enter homeownership. It is now common for mortgage loan products to offer low down payment options without private mortgage insurance. Private mortgage insurance (PMI) has been used by lenders in the past when the down payment is less than 20% of the purchase price. PMI protects the lender in the event that the borrower defaults on the loan or stops making mortgage payments. The insurance is typically charged as a premium paid with the monthly mortgage payment. The industry standard requires the borrower to pay the premium until 20% equity is reached in the home. The typical cost of private mortgage insurance ranges from 0.46% to 1.50% of the loan amount (Urban Institute, 2024). This additional amount can make homeownership even more difficult for prospective low-income homeowners. Due to rising housing costs, it is now common for mortgage financing terms to offer 3 to 5% down payment loans without private mortgage insurance. To offset the risk assumed by lenders, these loan terms may come with higher interest rates. However, the reduction in the amount required for the down payment provides entry into homeownership for households that lack the funds needed to cover the large sum of cash needed to close on a home.

There are also other mortgage options offered by federal government agencies that provide favorable financing terms for first-time and low-income homebuyers. Several examples include FHA, USDA, and VA loans. A Federal Housing Administration (FHA) loan is a mortgage issued by lenders that are insured by the government. FHA loans are often the most commonly available type of mortgage financing for first-time and lower income households. These loans require less income and accommodate lower credit scores, and therefore may provide mortgage finance for households that typically do not meet the underwriting requirements for conventional financing. Since these loans are riskier, FHA loans often have higher interest rates but may allow homebuyers that would otherwise not qualify for a mortgage to enter homeownership.

The United States Department of Agriculture (USDA) also offers mortgage financing for homeownership. Under the Single-Family Housing Guaranteed Loan Program, households in rural areas with household incomes below 115% of the area median income qualify for 100% financing, which requires no down payment. This funding can be used for the purchase, construction, or rehabilitation of a single-family dwelling unit used as the buyer's primary residence. Mortgage financing support is also provided by the U.S. Department of Veterans Affairs (VA) that supports veterans, servicemembers, and surviving spouses entering homeownership. VA loans require no down payment, no private mortgage insurance, and offer low interest rates and lower closing costs as a lifetime benefit each time a veteran (or another beneficiary) purchases a primary residence.

As the cost of entry to homeownership continues to rise, there is greater focus on innovative financing models. Several lending institutions now offer shared appreciation mortgage programs. In this model, the borrower shares a percentage of the home value's appreciation with the lender. In exchange for a portion of the appreciation, the lender offers below-market interest rates which reduces the monthly mortgage payments. This model is useful in high-cost markets where the average home is out of reach for moderate-income households. The lender is both a co-owner and investor in the property and takes a proportional stake of the gains or losses. Shared appreciation mortgages, which center on sharing future appreciation and allow homeowners to afford a home that is otherwise above their means, differ from shared equity mortgages that involve sharing the initial equity contribution and often help lower income borrowers enter homeownership (shared equity mortgages are described below in greater detail). In both cases, rising homeownership costs is the basis for driving the need for alternative financing to help make housing more affordable. There is growing concern about the potential predatory nature of these shared appreciation programs that has drawn the attention of many states.

Tax Credits

Tax credits can be used to incentivize entry into homeownership by reducing tax liability on a dollar-for-dollar basis. As an example, if a homeowner owes $2,500 in taxes, but has a $500 tax credit, their net tax liability drops to $2,000. In response to the financial crisis, a first-time homebuyer tax credit was adopted as part of the Housing and Economic Recovery Act of 2008. The program only lasted for a few years and expired for any homes purchased after May 1, 2010. Currently there is no federal tax credit for first-time homebuyers although there have been legislative attempts to revive one. The First-Time Homebuyer Act was introduced by Democratic Representative Earl Blumenauer of Oregon on April 28, 2021, but has not yet been passed into law. The bill would allow 10% of the purchase price of a home (up to $15,000) to be applied as a tax credit for households purchasing a primary residence that have not owned a residential property over the last three years.

Mortgage tax credit certificate programs, which are different than the mortgage interest deduction described earlier in this chapter, help first-time homebuyers and low-income households afford homeownership by providing tax credits tied to the mortgage. State housing finance agencies issue mortgage tax credit certificates to eligible homeowners to provide a dollar-for-dollar tax credit equal to a specified percentage of interest paid on the mortgage. The percentage varies state by state but is usually between 20 to 40% of the total mortgage interest. For example, a homeowner with a mortgage amount of $200,000 and interest rate of 4% with a 20% mortgage credit certificate would be able to claim up to a $1,600 tax credit on their annual federal income tax return. Since the program was established by the Deficit Reduction Act of 1984, state housing finance agencies have issued over 350,000 mortgage credit certificates, with more than half of all states participating in the program in 2020 (NCSHA, 2022).

Property taxes are a tax on real property assessed by local government to pay for public services such as education, law enforcement, and road construction. Many states and counties provide property tax credits and exemptions for low-income households, older adults, retirees, veterans, or persons with disabilities. The scale of the benefits provided by each program varies by jurisdiction. Generally, tax credits are provided if the property taxes exceed a certain percentage of the homeowner's gross income or if the household income is under a certain threshold. Many property tax credits for specific populations, such as veterans or persons with disabilities, cover a fixed percentage of the property tax bill. Some counties have a homestead tax credit for all households that use the property as their primary residence. It is more common for local jurisdictions to offer the homestead benefit as an exemption which reduces the assessed value that is taxed rather than a direct tax credit. The goal of tax credits and tax exemptions is to ensure that homeowners can stay in in their homes and not be forced to relocate as home prices – and the associated property taxes – rise.

Affordable Housing Types

Beyond assistance programs that make entry into homeownership more accessible, there are several housing types that are more affordable to purchase than single-family homes. Condos and townhomes fall within this category. Condos are units in a residential complex in which common areas are jointly owned by all owners, but each unit is owned individually. There is a homeowner's association that unit owners belong to, and unit owners pay regular fees into the association for the management and upkeep of common areas, building amenities, property exterior, insurance, and management. Although most housing markets have luxury condos in prime locations with extravagant views and amenities, the average condo housing stock tends to be more affordable than single-family homes. Condos are usually smaller units with less square footage. The costs associated with maintenance may be lower (depending on the age and construction quality of the building) since owners are only individually responsible for the interior of their unit, and the expenses for the building exterior and property grounds are shared. Homeowners insurance also tends to cost less since the building is covered by a master insurance policy that is shared by all owners. Amenities within the condo community (gyms, pools, party rooms, or business centers) allow residents to save costs on recreational memberships and meeting or event space. Condos also tend to be in areas that are more densely populated, such as cities, where basic services are within walking distance or accessible through public transportation, reducing the need for a personal vehicle and related car expenses.

Townhomes also provide greater affordability than many single-family homes. Like condos, townhomes are often smaller and, thus, priced lower. Unlike condos, townhomes do not always have homeowner associations. Some may have regular dues that help pay for shared expenses and this is typical in townhome developments with common areas. However, the lots may be subdivided and individually owned in some townhome developments. This can be true even with attached townhomes. If the townhomes are attached and there is no association, there is often an existing shared maintenance agreement that dictates changes and repairs to the exterior of the building that impacts connecting adjacent units. Townhomes offer a lifestyle of single-family living, while condo units are typically located in larger multi-family buildings.

In addition to condos and townhomes, smaller housing unit options are gaining traction as entry to homeownership has become increasingly unattainable for the average household. Micro units, micro houses, or tiny homes, at an average of 225 square feet in the U.S., offer a cheap homeownership option, especially for people living in expensive cities. This trend emphasizes location by sacrificing square footage and luxuries in private living spaces. This housing type is attractive for younger couples and single professionals who do not spend a lot of time at home.

A mix of these more affordable housing types can be collectively referred to as missing middle housing. The missing middle approach promotes forms of denser housing between single-family homes and large multi-family complexes. Examples include townhomes and rowhouses, duplexes/triplexes/fourplexes, cottages and bungalows, courtyard buildings, and live-work spaces. Missing middle housing can be built on developed parcels or using infill development on vacant lots, and these developments have lower minimum lot sizes, and provide a range in size and style of units. These housing options are desirable for millennials, young professionals, families with children, older adults, and multigenerational households. "Missing" is emphasized because for decades exclusionary zoning laws prohibited a diversity of housing types in single-family neighborhoods across America. In fact, New York is the only U.S. state in which single-family homes account for less than 50% of the total housing stock (Jones, 2022).

Communal Housing

Communal housing is an arrangement where a group of people live together in a community where each household has their own individual space but other areas in the community are shared, such as gardens, gyms, pool area, kitchens, and living spaces. This approach stands in stark contrast to the individually-focused living provided by condominiums. Furthermore, the form of ownership and financial structure associated with communal living differs from condos. A similar, but slightly different model is co-living that supports a similar lifestyle. Co-living is evident in assisted living facilities and student dorms, but these types of housing are usually not for sale. We highlight the two primary communal housing ownership types – cohousing and cooperatives – in the sections that follow.

Cohousing

The purpose of cohousing is to create a community among residents based on a set of common values. These values range from inclusive decision-making to how residents care for one another to sustainable living. Cohousing residents share and interact in community spaces but have their own individual living quarters. The development tends to be compact with smaller individual units and vehicles may be restricted to the periphery (no garages, roadways, or parking within the community) to encourage a communal environment. Cohousing development often incorporates design features that facilitate residential interaction such as a large common kitchen, gardens, indoor activity spaces, and outdoor gathering areas. A key component of the model is a set of shared responsibilities that makes child or elder care easier and enhances community for routine activities. For example, each resident may be responsible for cooking dinner for other residents on a regular basis, cleaning common areas, maintaining the landscaping, or driving other residents to their doctor appointments.

Although cohousing has been a form of housing in Europe since 1960s, it is relatively new in the U.S. The cohousing movement was inspired by architect Jan Gudmand-Hoyer of Denmark in an article titled, "The Missing Link between Utopia and the Dated One-Family Housing." The model has captured more attention as the need and desire for connection and a supportive network has grown in American society. Initially, the impetuous of cohousing was for multi-generational households with children to have additional family support. Today many smaller households without children and older adults are attracted to the cohousing model because it provides a sense of community and can help alleviate loneliness and isolation. In the U.S., the ownership structure for cohousing developments is often similar to condos, where each resident owns their own unit, pays their own taxes, and is responsible for a homeowner's association fee for shared spaces. This is largely because banks frequently offer financing options that are consistent with a typical condo-like structure. There are, however, exceptions where some cohousing units are structured like cooperatives which are discussed below.

Cooperatives

The lines between cohousing and cooperatives often blur since cohousing can be structured as a cooperative. And cooperatives are similar to cohousing developments because residents often have a set of shared responsibilities for the management and upkeep of the common areas. However, cooperatives differ in that community participation does not drive the model, rather the unique financing structure does. Cooperatives have an ownership structure where residents are shareholders of a corporation that hold title to the property. One's ownership share in the corporation is relative to the size of unit that they occupy. This structure is different than renting in a multi-family building because the ownership stake in the corporation can increase in value as the property appreciates. A cooperative, therefore, can function as a wealth building mechanism. There may be a mortgage on the development, but shareholders cannot get a typical mortgage for their specific unit. Rather, if the prospective shareholders need financing to buy shares in the corporation, they obtain a loan much like a mortgage but the shares, not the property, are used as collateral. Shareholders are also responsible for paying their pro-rata share of costs to maintain the property. In some cases, there may be a loan on the entire property which means each shareholder may be assessed fees related to the loan.

Cooperatives can be more affordable than condos. The shareholders can restrict resale prices to keep the property affordable which are known as limited equity co-ops and are a type of shared equity homeownership. Furthermore, since the management and maintenance of the property is often assumed by the shareholders, fees for services and general upkeep of the property are often reduced. For large capital improvements, the corporation may be able to use the property as collateral and borrow against the building instead of assessing each shareholder. If household composition changes, residents can exchange units to remain housed in that building instead of having to relocate to a different property when the household size or preferences change which helps keeps residents connected to their community.

Shared Equity Homeownership

Shared equity homeownership, also referred to as the third sector in housing, offers a middle ground between rental housing and market rate homeownership. As a tool to provide long-term affordable homeownership, shared equity homeownership programs provide an alternative to conventional homeownership by limiting potential capital gains through resale restrictions to make prices more attainable for lower income households. Resale restrictions ensure that the price of the home remains affordable when it is sold to the next owner. While there may be controversy surrounding resale restrictions since critics argue that wealth-building potential through property ownership should not be limited (Jacobus & Sheriff, 2009), shared equity programs provide homeownership opportunities for households that would otherwise be unable to purchase a home. Additionally, these programs still allow owners to accumulate wealth, although the price appreciation is capped (Acolin et al., 2021). Another advantage of shared equity homeownership is that homeowners are less likely to default on their mortgages and face foreclosure since the entity sharing the equity facilitates stewardship, which provides support to the household to enter favorable financing terms that the households can afford and provides assistance during times of hardship ensuring greater housing stability (Thaden, 2011). The popularity of shared equity homes is growing, with over 250,000 such homes existing in the United States as of 2018 (Thaden, 2018). There are four primary types of shared equity homeownership programs, including community land trusts, deed restricted/below market rate programs, limited equity cooperatives, and resident owned communities, also known as resident-owned manufactured communities.

Community Land Trusts

Community land trusts, also referred to as CLTs, first emerged in the United States about 50 years ago, but have been a popular model in other countries, primarily in Europe. There are now over 300 CLTs with more than 19,000 ownership homes in the United States (Wang et al., 2023). The primary key to affordability in this model is dual ownership.[1] In a CLT, a non-profit organization owns land and homes are built on this land. CLT residents may either rent housing from the trust or own their attachment to the property (their home). If the home is owned, the owner typically enters a 99-year land lease with the trust at a very affordable rate. Because CLT residents who purchase their home are only paying for the structure itself – not the land – CLT homebuying is more affordable. In addition, CLTs frequently provide subsidies to help finance the down payment and closing costs. CLTs also frequently have resale restrictions that limit the values at which residents can resell their home. These provisions ensure greater long-term affordability of these housing units. When the home is resold, the seller recoups their initial investment and a portion of the appreciation, but because of resale restrictions, prices may not reflect actual market rates for similar properties. On average, CLT shared equity homes are priced 31% lower than market rate (Wang et al., 2023). If the property's value decreases, the homeowner will not experience as large of a loss as they would with a market rate home, as the trust acts as a safeguard against market fluctuations and absorbs a portion of the loss as stipulated in the resale formula.

A key principle in the CLT model is stewardship. The nonprofit that owns the land ensures that CLT homes are maintained, remain affordable, and are transferred to income-eligible buyers that will use the home as their primary residence. Stewardship also applies to the relationship that the non-profit organization has with their residents. A primary goal is to ensure a positive homeownership experience for CLT residents. To support this goal, the non-profit may provide classes and training, help secure favorable financing terms with reasonable mortgage payments, and protect homeowners from foreclosure. The rate of foreclosure for CLT homeownership was 0.46% in a 2011 survey, compared to a rate of 4.63% among homeowners with conventional mortgages (Thaden, 2011). The stewardship activities during loan acquisition and post-purchase help to explain the low rates of delinquency and foreclosure in CLTs.

Deed Restricted/Below Market Rate Programs

In deed restricted homes, the resale restrictions are written directly in the deed so the property remains affordable to subsequent homeowners. The home may only be resold to buyers with qualifying incomes. For example, a deed restriction may limit future purchasers of the home to those with household incomes below 100% of the area median income. A deed will often include a formula to calculate the appropriate affordable purchase price for income-eligible buyers. All housing types such as single-family, townhomes, and condominiums can be deed restricted. The affordability period for these restrictions may be for a set time period or exist in perpetuity. The number of deed restricted homes is unknown in the United States. One of the first deed restricted homeownership program in the U.S. was established in Montgomery County, Maryland – the Moderately Priced Dwelling Unit program, which was established in the 1970s. By incorporating resale and occupancy restrictions in the deed, the Moderately Priced Dwelling Unit program mandates that a share of all residential units in the county are accessible to low- and moderate-income households. The program requires any housing project with 20 units or more to offer up to 12.5% of the units at affordable rates for 99 years for rentals and 30 years for ownership (Wong et al., 2021).

Limited Equity Cooperatives

Shared equity cooperatives generally function similar to cooperatives, where prospective residents purchase shares in the corporation, which gives them to the right to occupy a unit. The difference lies in how much equity a shareholder can capture. In a typical cooperative, shares change in price overtime similar to the stock market and the shareholder can buy or sell at any time. However, in shard equity cooperatives, the amount at which shares are bought or sold are controlled by the board. The provisions that dictate the amount of equity allowed are stated within the corporation by-laws which includes a formula that sets the price at which shares can be sold. This formula is designed to keep the shares affordable to prospective residents. There are about 200,000 limited equity units located in the nation, with the greatest concentration found in New York City (Urban Homesteading Assistance Board, 2015).

Resident Owned Manufactured Communities

Manufactured housing provides affordable homeownership for over 17 million Americans—roughly 6% of the housing stock—without a specific subsidy since they are cheaper to build (Prosperity Now, 2019). Manufactured homes are built to meet region-specific federal building standards as outlined in HUD's 1976 Manufactured Home Construction and Safety Standards. For years, manufactured housing communities have been at risk of redevelopment by for-profit developers as affordable land for new development becomes scarcer. In response, the resident-owned community model, also referred to as ROC, preserves manufactured housing as an affordable housing option that allows residents to remain in place. There is low-cost entry into the ROC, which functions like a community land trust; the land is owned and managed by the residents through either a non-profit or for-profit corporation, but the residents own their own manufactured home on the property. All residents belong to the corporation and elect a board of directors to manage the daily operations of the community. Resident-owned communities comprise about one third of the manufactured homes stock (Prosperity Now, 2019).

Manufactured homes in ROCs are a form of affordable homeownership for several reasons. First, manufactured homes are far cheaper to build, compared to site-built construction. Second, the ownership and management structure provide affordable and stable conditions through collective ownership. Last, when the resident-owned corporation has control of the land, the fee for the lot is stabilized. One downside to this model is that financing for manufactured homes often carries high costs because the home is titled as personal property rather than real property. This means that only chattel loans, which are loans that are used to purchase personal property, that have high interest rates are available instead of traditional mortgage financing.

Preserving Homeownership

Home Repair Programs

Home repairs can be a costly aspect of owning a home. Home repairs range from minor, and relatively affordable, fixes such as a leak under a sink, to large expenses such as the replacement of a roof. Older adults living on a fixed income and lower income homeowners often have insufficient funds to make such repairs. The inability to pay for home repairs leaves homes dilapidated and in disrepair which can create unsafe and unsanitary living environments. In order to help preserve homeownership, home repair programs have been established to help eligible households pay for the costs of home repairs in order to maintain the homes in good condition and to help households retain their home. These programs are offered by federal, state, and local governments as well as non-profit agencies. The form and amount of assistance varies by program but is usually offered in the form of a grant or loan. The money can only be used for certain types of repairs that improve safety hazards and housing quality as stipulated in the program. To be eligible, the home must be used as a primary residence and some programs have stipulations that require the owner to remain in the home for a certain period or they have to repay the money if they move out of the home quickly.

Property Tax Relief Programs

Besides mortgage payments, one of the largest reoccurring expenses of homeownership is property taxes. Older adults on fixed incomes and lower-income homeowners in gentrifying neighborhoods are particularly vulnerable to losing their homes due to defaulting on property tax payments. Inflation and rapid home appreciation drive increases in property taxes since these taxes are calculated based on the assessed value of the home. County assessors establish assessed values based on local housing market conditions. Planning for property tax payments can be challenging, especially in a location where home prices are increasing at a rapid, and unpredictable, rate. Many states offer tax relief programs based on a range of eligibility criteria. It is most common for states to reduce the assessed value of the property by a certain amount or percentage, which is often used as the base to calculate the tax amount. A less common method, referred to as a circuit breaker, is to "shut off" taxes once the tax amount exceeds a certain portion of the household's income. Common eligibility criteria are based on income, age, veteran or disability status, and the home's use as a primary residence. Although not as common, some states have tax exemptions or credits for widows and people in certain occupations (e.g., government employees, first responders, clergy). Disaster damage or homes located on certain types of land uses such as agricultural land or near land uses that are less desirable also receive tax exemptions or credits in a few states.

Foreclosure Prevention Initiatives

Lower income, non-white households have been hit hardest by foreclosures, especially during the Great Recession due to non-traditional forms of mortgage financing (Garriga et al., 2017). Borrowers of color are more than twice as likely to experience foreclosure than white homeowners and foreclosure rates are highest for low-income borrowers (Bocian et al., 2011). Foreclosure prevention is key to reducing housing instability and allowing households to remain in place. In addition to the obvious benefits of residential stability, avoiding foreclosure also allows households to keep the equity (if any) they have in their home and continue to build wealth over a longer period of time. Foreclosure prevention programs help households at risk of default and those already behind on their mortgage payments keep their homes.

There are a wide range of assistance programs available, including mediation and legal advice, refinancing, loan modification, and emergency financial assistance. Mediation is frequently offered by states or local government to homeowners to explore possible alternatives to foreclosure. This is different than working with a lawyer for legal advice, as mediators take a neutral stance and seek to support lenders and homeowners to come to a mutually agreeable resolution. During refinancing, a homeowner replaces their existing mortgage with a new one, which usually provides financial advantages, such as a lower monthly payment. A homeowner may also refinance in order to change the term of the loan; a shorter mortgage term allows the homeowner to pay off the loan more quickly.

Additional strategies to prevent foreclosure include loan modification, deferred mortgage payments, and changing the terms of a mortgage loan. This is different than refinancing, as the entire mortgage is not being replaced with a new one. Last, state and local governments, often in partnership with non-profit organizations, provide emergency financial assistance to income-qualifying homeowners that are at-risk of or have defaulted on their mortgage payments. Together, these range of initiatives help to prevent or limit foreclosures for lower income households.

Conclusion

While much of the attention related to affordable housing focuses on renting, there are a range of tools to provide access to homeownership for lower income households. In this domain, there are strong differences of opinion about whether homeownership is the proper mechanism for achieving housing affordability. On one side, are those who believe scarce resources should be used to subsidize the development of affordable rental housing, or provide direct subsidies to low-income households that can be used to rent housing. The belief being that the reliance on homeownership in the United States has helped create the affordable housing crisis in the nation. Why, goes the argument, would we perpetuate this flawed policy by encouraging households with limited financial resources to purchase their own home?

The other side of the argument acknowledges that, for better or worse, homeownership is a primary mechanism for generating household wealth in the United States. Therefore, low-income households and households of color who have historically had limited access to homeownership have not had the opportunity to generate wealth. As a result, steps should be taken to remedy this lack of opportunity by providing access to homeownership to households that have been historically excluded from this opportunity. The affordable homeownership approaches outlined in this chapter demonstrate that many creative and successful programs have been established and continue to provide access to homeownership for lower income households. Despite these efforts, there remains a significant gap in rates of homeownership along socioeconomic and racial lines. Therefore, if the nation determines that homeownership should be emphasized, the scale of these programs and access to them will need to grow.

Note

1 Community Land Trusts can also provide rental housing to tenants. The trust owns the home and then leases the home to households who do not have any ownership stake in their housing unit.

References

Acolin, A., Goodman, L., & Wachter, S. M. (2016). A renter or homeowner nation? *Cityscape: A Journal of Policy Development and Research*, *18*(1), 145–157.

Acolin, A., Ramiller, A., Walter, R. J., Thompson, S., & Wang, R. (2021). Transitioning to homeownership: Asset building for low- and moderate-income households. *Housing Policy Debate*, *31*(6), 1032–1049.

Board of Governors of the Federal Reserve System. (2023). *Changes in U.S. family finances from 2019 to 2022. Evidence from the survey of consumer finances*. Retrieved from www.federalreserve.gov/publications/files/scf23.pdf

Bocian, D. G., Li, W., Reid, C., & Quercia, R. G. (2011). *Lost ground, 2011: Disparities in mortgage lending and foreclosures*. Center for Responsible Lending. Retrieved from www.responsiblelending.org/mortgage-lending/research-analysis/Lost-Ground-2011.pdf

Choi, J. H., Goodman, L., & Pang, D. (2024). *Rent reporting can help build credit. Why aren't smaller-property tenants opting in?* Retrieved from www.urban.org/urban-wire/rent-reporting-can-help-build-credit-why-arent-smaller-property-tenants-opting

Congressional Budget Office. (2021). *The distribution of major tax expenditures in 2019*. Retrieved from www.cbo.gov/publication/57585.

Dey, J., & Brown, L. M. (2020). The role of credit attributes in explaining the homeownership gap between whites and minorities since the financial crisis, 2012–2018. *Housing Policy Debate*, *32*(2), 275–336.

Einstein, K. L., Ornstein, J. T., & Palmer, M. (2023). Who represents the renters? *Housing Policy Debate*, *33*(6), 1554–1568.

Fischer, W., & Sard, B. (2017). *Chart book: Federal housing spending is poorly matched to need. Tilt toward well-off homeowners leaves struggling low-income renters without help*. Washington, DC: Center on Budget and Policy Priorities.

Garriga, C., Ricketts, L, R., & Schlagenhauf, D. E. (2017). The homeownership experience of minorities during the great recession. *Federal Reserve Bank of St. Louis Review*, First Quarter 2017, *99*(1), 139–167.

Goodman, L., & Zhu, J. (2018). *Rental pay history should be used to assess the creditworthiness of mortgage borrowers*. Retrieved from www.urban.org/urban-wire/rental-pay-history-should-be-used-assess-creditworthiness-mortgage-borrowers

Hanson, A. (2012). Size of home, homeownership, and the mortgage interest deduction. *Journal of Housing Economics*, *21*(3), 195–210.

Institute on Assets and Social Policy (IASP) & National Low Income Housing Coalition. (2017). *Misdirected investments: How the mortgage interest deduction drives inequality and the racial wealth gap*. Retrieved from https://nlihc.org/sites/default/files/MID-Report_0817.pdfhttp%3A/NLIHC-IASP_MID-Report.pdf

Jacobus, R., & Sherriff, R. (2009). *Balancing durable affordability and wealth creation: Responding to concerns about shared equity homeownership*. Washington, DC: Center for Housing Policy.

Jones, J. (2022). *U.S. cities with the most single-family homes. Construction coverage*. Retrieved from https://constructioncoverage.com/research/cities-with-the-most-single-family-homes-2021

Lautz, J. (2022). *Tackling home financing and down payment misconceptions*. National Association of Realtors. Retrieved from www.nar.realtor/blogs/economists-outlook/tackling-home-financing-and-down-payment-misconceptions.

McCabe, B. J. (2016). *No place like home: Wealth, community and the politics of homeownership*. New York: Oxford University Press.

McCabe, B. J. (2018). Costly, regressive, and ineffective: How sensitive is public support for the mortgage interest deduction in the United States? *Housing Policy Debate, 28*(6), 963–978.

Mills, G., Gale, W. G., Patterson, R., Engelhardt, G. V., Eriksen, M. D., & Apostolov, E. (2008). Effects of individual development accounts on asset purchases and saving behavior: Evidence from a controlled experiment. *Journal of Public Economics, 92*(5–6), 1509–1530.

NCSHA. (2022). *Mortgage credit certificate program Q&A*. National Coalition of State Housing Agencies (NCSHA). Retrieved from www.ncsha.org/resource/mortgage-credit-certificate-program-qa/

Perez, L. (2022). *Average U.S. savings account balance: A demographic breakdown*. ValuePenguin | Lendingtree. Retrieved from www.valuepenguin.com/banking/average-savings-account-balance.

Pradhan, A. (2022). *Average total down payment reached an all-time high in 2022*. CoreLogic. Retrieved from www.corelogic.com/intelligence/average-total-down-payment-reached-an-all-time-high-in-2022/

Prosperity Now. (2019). *Facts about manufactured housing – 2019*. Retrieved from https://prosperitynow.org/resources/facts-about-manufactured-housing-2019

Scally, C. P. (2012). The nuances of NIMBY: Context and perceptions of affordable rental housing development. *Urban Affairs Review, 49*(5), 718–747.

Sommer, K., & Sullivan, P. (2017). Implications of U.S. tax policy for house prices, rents, and homeownership. *American Economic Review, 108*(2), 241–274.

Teles, D., Ramos, K., Su, Y., & Su, D. (2023). *Using vouchers to support homeownership*. Retrieved from www.urban.org/research/publication/using-vouchers-support-homeownership#:~:text=The%20HCV%20homeownership%20program%20is,11%2C672%20active%20vouchers%20in%202021.

Thaden, E. (2011). *Stable home ownership in a turbulent economy: Delinquencies and foreclosures remain low in community land trusts*. Working paper. Cambridge, MA: Lincoln Institute of Land Policy. Retrieved from www.jstor.org/stable/pdf/resrep18350.1.pdf

Thaden, E. (2018). *The state of shared equity homeownership*. Shelterforce. May 7. Retrieved from https://shelterforce.org/2018/05/07/shared equity/.

Thaden, E., & Wang, R. (2017). *Inclusionary housing in the United States: Prevalence, impact, and practices*. Working paper. Cambridge, MA: Lincoln Institute of Land Policy. Retrieved from www.lincolninst.edu/publications/working-papers/inclusionary-housing-united-states

Urban Homesteading Assistance Board. (2015). *Building capacity to serve and grow the cooperative housing community*. Retrieved from www.uhab.org/wp-content/uploads/2021/01/Research-Update-12.9.15.pdf

Urban Institute. (2024). *Housing finance at a glance. A monthly chartbook*. Retrieved from www.urban.org/sites/default/files/2024-03/Housing_Finance_At_A_Glance_Monthly_Chartbook_March_2024.pdf

Ventry, D. J. (2009). The accidental deduction: A history and critique of the tax subsidy for mortgage interest. *Law and Contemporary Problems, 73*, 233–284.

Wang, R., Wandio, C., Bennett, A., Spicer, J., Corugedo, S., & Thaden, E. (2023). *The 2022 census of community land trusts and shared equity entities in the United States: Prevalence, practice, and impact*. Working Paper WP23RW1. Cambridge, MA: Lincoln Institute of Land Policy.

Wachter, S., & Acolin, A. (2022). Homeownership for the long run. *Journal of Comparative Urban Law and Policy, 5*(1), Article 23, 274–296.

Woodward, S. E. (2008). *A study of closing costs for FHA mortgages*. Washington, DC: U.S. Department of Housing and Urban Development.

Wong, B., Lung-Amam, W., & Knapp, G. (2021). *Moderately priced dwelling units: Montgomery County, Maryland's model of inclusionary housing*. Cambridge, MA: Lincoln Institute of Land Policy.

Chapter 11

Regulatory Strategies

Overview

Government can promote increased affordability of housing by implementing a variety of policy tools that have been described in the preceding chapters. Governments – at all levels – can develop publicly-owned housing, issue tax credits to affordable housing developers, and provide housing vouchers to eligible households. For most readers of this book, these are the programs that most likely come to mind when people think of government involvement in affordable housing. But there are an entirely different class of actions that governments can pursue. In this chapter, we highlight how regulation – or a lack thereof – can influence the supply of affordable housing. Unlike more well-known housing policies, regulatory strategies frequently do not require public funds to implement. Rather, the primary governmental *costs* of regulatory programs are typically political, rather than financial. The following sections highlight the different regulatory levers that governments can pull to provide greater housing access and affordability.

Federal Regulatory Mechanisms

Lower-income households, and households of color, have struggled to find safe, decent, and affordable housing not only because they are often priced out of the housing market, but also because they have frequently been subject to housing market discrimination. In response, the federal government has taken steps to respond to these challenges. The two most notable federal regulations designed to address these forms of discrimination are the Fair Housing Act and the Community Reinvestment Act. Both seek to end discrimination in the housing market and reduce high levels of residential segregation.

The Fair Housing Act

The Fair Housing Act, also referred to as Title VIII of the Civil Rights Act of 1968, is designed to protect individuals from discrimination in the housing market. Discrimination based on race, national origin, gender, religion, disability, or familial status is prohibited under the act. These categories of individuals are referred to as protected classes under the law. This protection extends to renting a home, buying a home, obtaining a mortgage, and all other housing-related activities. The only housing types that aren't covered by this act are buildings with less than five units that are owner-occupied, housing owned by religious organizations, and private clubs that are limited to members. For example, if a homeowner of a duplex resides at that duplex as their primary residence and leases out the adjoining unit, the Fair Housing Act does not apply. In

DOI: 10.1201/9781003356585-14

that case, an owner could reserve the leased unit exclusively for female applicants. However, if the owner does not reside in the duplex, the Fair Housing Act applies, and the owner could not restrict access to the unit based on gender during the rental process.

As discussed in Chapter 5, residential segregation and persistent discrimination has made the Fair Housing Act a vital regulatory mechanism for equitable treatment of all individuals in the housing market. The primary goal of the Act was to limit discrimination in the housing market. After passage of the Fair Housing Act, it became illegal for homeowners to refuse to sell a home to a Hispanic household, require different terms and conditions for the sale of the home to an Asian household, or establish a different sales price for a Black household. Mortgage lenders cannot refuse to provide loans or impose different loan terms or conditions for members of a protected class. If this, or any type of discrimination occurs in the renting or purchase of a home, a formal complaint can be filed with U.S. Department of Housing and Urban Development (HUD). If the investigation indicates a possible violation of law, the case may be handled by the U.S. Department of Justice or HUD's Office of General Counsel. An anti-discrimination law, such as the Fair Housing Act, is an important first step in ending housing discrimination, but enforcement of such laws is equally important. One of the challenges of regulations designed to address discrimination is the issue of enforcement. The primary issue is that the burden falls on the victim to report the violation and go through the lengthy complaint process. For households that lack the time and resources to initiate and participate in such a process, this is a heavy burden. Further complicating matters is the fact that acts of discrimination are frequently difficult to prove. In sum, the existence of anti-discrimination laws is necessary, but insufficient to address discrimination. Effective enforcement – easier said than done – is also necessary.

Eliminating outright discrimination in the housing market was the primary goal of the Fair Housing Act. A secondary goal was to reduce high levels of residential segregation (as discussed in Chapter 5). The law was designed to prevent discrimination, with the hope that by decreasing instances of overt discrimination, a secondary byproduct could be lower levels of segregation. As Katherine O'Regan wrote, the law had fairly good success reducing outright discrimination, but its impacts on levels of segregation are muted at best (O'Regan, 2018).

One of the features of the Fair Housing Act that sought to reduce residential segregation was the concept of affirmatively furthering fair housing (AFFH). AFFH was designed to reverse patterns of residential segregation by requiring recipients of federal financial assistance to take action to mitigate persistent patterns of segregation, discrimination, and disinvestment. For many years, the AFFH obligation was realized by requiring HUD recipients – cities, counties, and local housing authorities – to produce fair housing plans every five years. These plans had to identify impediments to fair housing, create a list of actions to overcome the impediments, and maintain records that documented the actions that were taken. This process is referred to as an analysis of impediments to fair housing choice, also known as the AI. In reality, oversight of the AI was limited and there was little enforcement. As a result, the AFFH rules had limited impact in the first few decades of the Fair Housing Act.

To enhance the impact and effectiveness of AFFH, the Obama Administration established the 2015 AFFH rule and required local jurisdictions and housing authorities receiving HUD funding to make concrete commitments, and publicly track and report their mitigation efforts with a defined set of data and metrics. These rules on enforcing AFFH have gone through many revisions and are subject to the political whims of the political party in charge. In 2020, the Trump Administration repealed the AFFH rule by deeming it unnecessarily burdensome. In its place, it established the Preserving Community and Neighborhood Choice (PCNC) rule. The PCNC rule required recipients of HUD funding to certify that they will promote fair housing but there were no specific actions, assessment, or reporting required. Rather, it simply required written commitment which significantly relaxed the obligations of recipient entities. Many scholars and advocates saw the repeal of AFFH as a step backward in efforts to mitigate and end housing discrimination and residential segregation (NLIHC, 2020).

After the Biden Administration took office, part of the 2015 AFFH rule was restored. The new AFFH interim final rule established in 2021, referred to as AFFH IFR, rescinded the PCNC rule and required recipients to make meaningful progress toward affirmatively furthering fair housing in their planning activities through consolidated plans, annual action plans, and public housing authority plans. In 2023, AFFH went through another revision to improve accountability by requiring HUD recipients to conduct fair housing analyses, develop an equity plan, outline specific actions to address fair housing, and submit annual progress reports. Public filing of the information was established to allow the public to hold HUD recipients accountable. The intent was that the stricter requirements will help reverse historic patterns of residential segregation and reduce or eliminate discrimination in housing-related activities. The goal is to foster inclusive communities and provide equal access to safe, decent, and affordable housing for all.

Community Reinvestment Act

The Community Reinvestment Act (CRA) was passed in 1977 with two specific goals: (i) to prevent discriminatory lending practices, and (ii) to increase credit and deposit opportunities to low-income households and small businesses in under-resourced neighborhoods. Regulated financial institutions, like Bank of America and JP Morgan, are required to serve the lending and depository needs of the communities they serve. Each institution defines an assessment area which aligns with the geographic areas they serve based on the location of their bank branches and deposit-taking ATMs. The Federal Deposit Insurance Corporation (FDIC), a federal regulator, provides oversight and assesses the records of the financial institutions to ensure they comply with the act. The Federal Reserve system and Office of the Comptroller of the Currency also provided oversight. The CRA assessment includes three different tests: a lending, investment, and service performance review. The lending test examines the number and amount of loans originated in low-income census tracts within the bank's assessment area. The investment test assesses the dollar amount of qualified investments and how flexible lending practices address the needs of low-income individuals. The service test evaluates the distribution of bank branches in their service area and the accessibility of banking services to low-income households. For each of these tests, one of five ratings is assigned – Outstanding, High Satisfactory, Low Satisfactory, Needs to Improve, and Substantial Noncompliance – and the scores are made public.

The procedures for CRA assessment depend on the lending institution. Large banks are the only entities that must comply with the full list of CRA requirements highlighted in the previous paragraph. Smaller banks, for example, have a more streamlined set of requirements, while limited purpose and wholesale banks only need to comply with community development needs (e.g., affordable housing, health or social services, services for low-income families) through the combination of lending, investment, and services. Banks that receive an outstanding rating go through the process every five years and those with satisfactory ratings are assessed every four years. At any time, the FDIC can conduct a review for cause. Banks have a strong incentive to comply with the CRA, because there are significant business implications. In addition to the reputational risk associated with failure to comply with the CRA, the scores can impact regulatory approval of long-term strategies such as opening or closing of branch offices, purchases of or mergers with other institutions, or changes to the bank's business operations.

A significant focus of research has been the impacts and effectiveness of the CRA and the results have been mixed. On one hand, the CRA has led to $1.7 trillion of new loans in economically distressed areas and has increased access to credit for households of color (Friedman & Squires, 2005). However, critics argue that CRA does not go far enough to help moderate- and low-income borrowers. Regulators have been accused of being too lenient when faced with banks that have not made significant efforts to comply with CRA. The high ratings that banks frequently receive have been questioned because satisfactory scores can still be earned despite being noncompliant in some assessment categories. To strengthen oversight and to make compliance more stringent, observers have suggested a range of reforms: subject all lending institutions to an examination, require affiliate lending activities to be included in the evaluation, factor race-based lending patterns into evaluations, and redesign quantitative metrics and benchmarks standards for CRA activities (Goodman et al., 2022; Reid, 2019).

A balanced assessment of the Fair Housing Act and the CRA would suggest that these federal regulations have had a positive impact on limiting housing discrimination and increasing housing access. But these positive impacts may not be sufficient to deem both laws successful. Ample research highlights the shortcomings of both laws and underscores the unmet hopes and promises of these landmark policies.

Local Regulatory Mechanisms to Produce Affordable Housing through Residential Densification

The vast majority of regulations that govern housing construction and the housing market are established at the state or local level. Residential densification – a common approach to increasing housing supply and reducing costs – can, in part, be achieved by changing existing regulations. Existing land use regulations and zoning that prohibit all housing other than single-family housing are obvious impediments to greater residential density. Single-family zoning is the dominant form of land use in most metropolitan areas in the United States. Examples of regulatory changes include zoning reforms that expand land use from just single-family homes to uses that allow multi-family structures, inclusionary zoning/housing that incorporates affordable housing units in new developments, and a range of incentives that encourage infill housing development.

Zoning Reform

Local governments assign a set of regulations, called zoning, to real property to dictate its use. For example, grocery stores, office buildings, and manufacturing are not allowed in areas zoned for residential use. Broad zoning categories are residential, commercial, and industrial. The first zoning laws in the United States were established in the early 1900s in Los Angeles and New York. One of the purposes of zoning was to segregate industrial and commercial activity from residential neighborhoods. Because of the pollution and health hazards of industry, zoning was a way to organize a city and to address public health concerns. Zoning was also used for less virtuous purposes; these rules were also used to segregate people based on class or race. The history of single-family zoning in the U.S. has an explicitly racist history that has helped to perpetuate

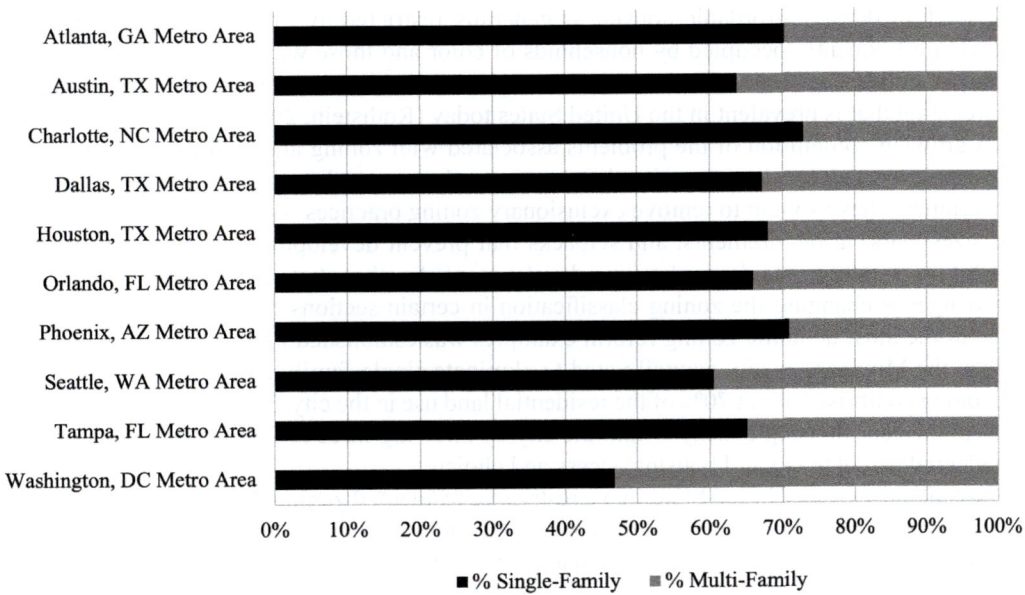

Figure 11.1 Percentage of Single-Family and Multi-Family Housing Units in the Ten Fastest Growing Metropolitan Statistical Areas (MSAs), 2016–2021

(Source: American Housing Survey, U.S. Census Bureau)

the significant residential segregation that is evident in many U.S. cities today. Another negative byproduct of the excessive use of single-family zoning is that it has inhibited the growth of the housing stock. As communities have grown, the ability to construct sufficient housing for the growing population has been constrained by restrictive zoning and land use.

How land can be used – its zoning – has a clear impact on the affordability and availability of housing. For much of the last century, exclusionary zoning practices have hindered the development of the diverse housing types that are required for a healthy housing system. Rather, in many jurisdictions, local governments have prohibited the construction of any type of housing other than single-family homes. Single-family zoning is very prominent in many cities; approximately 70% of the land use in most cities is designated for detached homes only (Badger & Bui, 2019). Figure 11.1 lists the top ten fastest growing metropolitan statistical areas (MSAs) in the U.S. from 2016–2021 and highlights the breakdown of their housing stock in 2021. With the exception of Washington DC, all of these MSAs are dominated by single-family housing, a significant constraint when facing rapid population growth.

This regulatory approach has had profoundly negative consequences for the nation and its households, particularly households of color and those with lower levels of income. The problem with exclusionary zoning is that it prevents higher density housing developments that can increase the number of units of housing on each parcel of land. Without sufficient multi-family housing, many households may struggle to find housing that meets their specific needs or obtain housing that they can afford. Detached single-family housing works well for millions of households in the country, but it doesn't work for everyone.

Exclusionary land use practices have also contributed to greater residential segregation (Gyourko & Molloy, 2015). When multi-family housing is prohibited in a majority of neighborhoods, the stock of denser, affordable housing is generally restricted to a few select neighborhoods, which are frequently occupied by households of color and those with lower levels of income. This pattern has been repeated over and over and has, in part, been responsible for the residential segregation that is prevalent in the United States today (Rothstein, 2017).

A growing recognition of the problems associated with zoning and land use have prompted governments at all levels to examine this practice and propose alternatives. In some cases, local governments have sought to remove exclusionary zoning practices – such as required minimum lot sizes, parking requirements, and setbacks that prevent development from being built near the lot line – to promote denser housing development. Another strategy to address affordability challenges is changing the zoning classification in certain sections of a city, or an entire city. One of the most dramatic zoning reform examples was established in the City of Minneapolis. When the Minneapolis city council voted to eliminate single-family zoning in 2018, that form of zoning comprised about 70% of the residential land use in the city. The purpose of this reform was to increase housing density within the city by allowing three to four housing units on each residential parcel to expand housing access and choice.

Not all cities in the United States have restrictive residential zoning. Houston, for example, is unique as it is the only large city in the country that does not have zoning regulations. If you've visited Houston, you may have noticed that the built environment differs from most other cities in the nation. Residential and commercial activity is blended throughout the city, whereas in most other places those functions would be physically separated by land use ordinances. From a housing standpoint, Houston has significant advantages due to this lack of zoning. It is easier to build, and housing can be built more quickly. These factors, at least in part, help to explain the greater housing affordability observed in Houston as compared to many other cities in the nation. But despite the ease of construction, the availability of various housing options may not be guaranteed. Single-family housing continues to dominate the housing market in Houston.

Given the obvious benefits associated with land use reform, readers may wonder why single-family zoning remains dominant across many jurisdictions in the United States. Abundant research has highlighted the significant challenges that exist when attempting to change land use. Efforts to eliminate single-family zoning are frequently opposed by incumbent residents who worry about changes to the character of their neighborhood, a decline in home values, the stress on existing infrastructure, traffic and parking problems, and overcrowding in schools (Einstein et al., 2019). Local land use battles are at the political forefront in the national conversation about affordable housing.

While there is strong support for zoning reforms, these actions alone may not be sufficient to increase housing supply and promote affordability. In Minneapolis, the new zoning reforms did not produce an immediate increase in multi-family housing. The reason for the delayed response was the city's failure to change other building standards in tandem with the elimination of single-family zoning (Lee, 2022). For example, a triplex that could now be constructed under the new land use policy was required to fit into the same height and floor area requirements that had previously existed when the land was zoned single-family. The result was a relatively modest increase in multi-family housing in the first couple years after the change in land use (Britschgi, 2022). Subsequent regulatory changes in Minneapolis have been implemented to accelerate the development of multi-family housing but it is too early to tell if the changes have had any positive impact yet. As the case of Minneapolis demonstrates, eliminating single-family zoning is not a silver bullet, but increasing the housing stock without such changes to land use will be nearly impossible. Therefore, for zoning changes to have the most significant impact, they must be accompanied by other regulatory changes to parking minimums, maximum building height, floor area ratios, and lot setbacks.

An interesting development in the movement for significant land use reform is the ways in which different levels of government have been involved. Because of the potential benefits associated with land use reforms (including the conspicuous social and racial inequities), the federal government has taken steps to promote zoning reform through grants and incentives. "Yes In My Backyard" (YIMBY) is a new federal program offered by HUD that provides grants to local jurisdictions and regional planning organizations that seek to increase housing production. Several types of eligible land use reforms that this program supports include minimum lot size reductions, expedited permitting, relaxed parking requirements, and incentives for accessory dwelling units.

There is also increased focus on zoning and land use by state governments. Because so many local jurisdictions have failed to make the land use changes necessary to accommodate increased housing production, states are now taking preemptive action to promote or, in some cases, require changes to land use. Oregon requires municipalities with populations greater than 10,000 to allow duplexes in areas previously zoned for single-family. California allows multi-family housing on commercially zoned properties and has reduced parking requirements near public transit. Connecticut prohibits caps on the number of multi-family units in a development and minimum floor area requirements for housing units. These new state standards help align policies to promote a regional approach to address the housing affordability crisis and promote change that may have previously been frustrated by local political opposition to fundamental regulatory change.

Inclusionary Housing

Inclusionary housing, or inclusionary zoning, is a regulatory mechanism used by state and local governments to encourage, or require, inclusion of affordable housing in new residential development. Private developers are encouraged (voluntary programs) or required (mandatory programs) to increase the affordable housing stock by designating a certain proportion of the units in a development for households at or below the area median income. Developers are also frequently provided an opt-out in which they pay a fee in lieu of providing the affordable units. Those fees are then used to produce affordable units at a different location. It is important to note that the opt-out is often criticized since it locates affordable housing in other areas of the city that may not be as desirable, delays affordable housing development, or the funds may not be used directly for affordable housing development. Although the requirements of each program vary, there are common elements found in all inclusionary housing programs. The requirements stipulate the minimum size of development that must comply (new developments with at least 10 units, for example) and the percent of units that need to be affordable (i.e., 15% of all units must be affordable). Often incentives are provided to developers to offset the costs of turning market-rate units into affordable housing. These incentives may include expanded land use, expedited permitting to fast track the project through the project review and permitting phase, fee waivers for the costs associated with the development review process, and relaxed site regulations such as reduced parking requirements or adjustments to height limits.

There are arguments for and against inclusionary housing programs and research supports both sides of this argument. Opponents cite research that shows that inclusionary housing programs increase housing costs by constraining supply while proponents highlight the number of new affordable units produced as a result of these programs (see Phillips 2024 for a more detailed discussion of the evidence and literature on this topic). The evidence is fairly clear that inclusionary zoning policies – at some point – have a negative impact on housing production. For example, in 2023, the City of San Francisco loosened its inclusionary zoning requirements in recognition of this negative impact. Under the revised rule, the city reduced the required number of affordable units in multi-family developments from 21.5% to between 12 and 16%.[1] Note, San Francisco did not terminate the program, rather it lessened the affordable housing requirements. However, existing stock of affordable units – that were created pursuant to inclusionary zoning – is a key marker of success of the program.[2] There are now over 1,000 inclusionary housing programs in the United States that appear in more than half of the states (Wang & Balachandran, 2021). The rise in inclusionary housing programs and emphasis on data collection will allow future research to further uncover the impact of these programs on housing supply and affordability.

Infill Housing

Infill housing involves developing new housing in areas that are already fully (or close to fully) developed. Infill development is a way to increase density and housing opportunities without resorting to urban sprawl. Infill housing may or may not be affordable but helps increase the overall housing stock. It benefits communities by reinvesting in abandoned buildings and underutilized or vacant lots. In addition to taking advantage of underutilized parcels of land, infill development has other advantages for communities. First, infill development has obvious environmental benefits. By constructing denser housing in the urban core, this development can provide additional housing options for urban dwellers that may not require long commutes from sprawling suburbs. Second, the additional supply of housing – particularly in denser developments – can provide affordability benefits. Rather than constructing single-family homes in the outskirts of a metropolitan area, dense, infill development can create far more units and help to bring much needed supply to tight housing markets. Such contributions to supply can reduce housing costs or prevent more significant increases in the cost of housing.

Communities may employ infill housing strategies on a lot-by-lot basis or more comprehensively using an approach such as land banking. Local governments use land banking to acquire multiple parcels to develop affordable housing. Once the land is under the control of a public or non-profit entity, the housing may be constructed more affordably and, as a result, the operator can charge lower rents. Control of land is a critically important tool in the fight to create and preserve affordable housing.

Infill housing can, in certain circumstances, be difficult because of a host of regulatory barriers. As an example, accessory dwelling units have historically been illegal in many cities. This is starting to change as many local governments have recently changed zoning regulations to permit smaller housing units to be developed on single-family lots. Furthermore, it is often difficult to reach economies of scale as most infill housing projects are small. This small-scale development can exacerbate already high development costs associated with construction costs in the urban core since overhead and development costs per unit are higher for smaller projects. Infill development may also receive push-back from "Not In My Backyard" (NIMBY) neighbors who are concerned about traffic congestion, lack of parking, increased noise, and changes in the character of urban neighborhoods. While these barriers can, at times, be significant, they should not deter efforts to develop dense, infill residential developments in the urban core.

Local Regulatory Mechanisms to Produce Affordable Housing through Development Provisions

In addition to strategies employed by local jurisdictions to promote denser residential development (as described in the prior section), states and local governments also use regulatory mechanisms to promote the construction of affordable housing. This outcome can be achieved through a range of regulations, particularly when they are used together, that affect the development process. Examples of such regulatory changes include, expedited permitting, linkage and impact fees, and community benefit agreements.

Expedited Permitting

All new residential development and major renovations are reviewed by the local jurisdiction to ensure development and construction complies with land use, building codes, environmental reviews, and all other regulations. At the conclusion of this review process, a permit is issued to the developer or contractor to allow the project to start. This approval can be lengthy and any delays can be costly for the developer. Additional costs associated with a slow permitting process have a direct impact on the cost of a project, and thus, the rents that will ultimately be charged. Expedited permitting is a tactic used in some jurisdictions to accelerate the approval process for projects that meet certain criteria, such as the development of affordable housing. The logic is that lengthy permitting processes increase the cost for the developer which is counter to the goal of delivering affordable housing. This strategy doesn't necessarily create new affordable housing development but rather allows for faster delivery of affordable housing at lower costs.

Expedited permitting for affordable housing may cover a range of activities, including environmental reviews, building permits, land use changes, rezoning applications, and requests for variances (e.g., building closer to the lot line than regulations allow or reducing parking requirements). Expedited permitting may be offered as a separate fast-track process, or a jurisdiction can simply prioritize affordable housing projects within the existing process. In some cases, affordable housing projects are provided early support and assistance in which staff from the local jurisdiction conduct a preliminary review and help the applicant solve any problems before the official application is submitted.

There are several factors to consider when designing an expedited permitting program. First, rules must clarify who qualifies as an eligible applicant. For example, what income levels does the project need to serve to be considered affordable? How many units in the development must be affordable? How long do the units need to remain affordable? Second, jurisdictions must clarify the geographic boundaries in which expedited permitting is available. Will development in all neighborhoods be eligible for expedited permitting or only certain neighborhoods within the municipality? If local officials want to spur development in particular areas, there may be defined boundaries for the program. Third, local jurisdictions may waive certain requirements if a project does not risk the health and safety of residents in a community.

A recent example of expedited permitting is Executive Directive 1 (ED 1) in Los Angeles. ED 1 was one of Mayor Karen Bass' first initiatives focused on housing. According to the city's description, ED 1 "expedites the processing of shelters and 100% affordable projects in Los Angeles. Eligible projects receive expedited processing, clearances, and approvals through the ED1 Ministerial Approval Process." ED 1 established a different review process for affordable housing projects and directs "all applicable City Departments to process clearances and utility releases related to building permit applications, certificates of occupancy, or temporary certificates of occupancy within 5 business days for 100% affordable projects and within 2 business days for Shelters."[3] Early evidence suggests that ED 1 combined with state density bonuses for 100% affordable developments have allowed private developers to construct affordable housing without public subsidy (Christopher, 2024). This move by Mayor Bass highlights the important role that regulation can play in promoting the construction of much needed affordable housing.

Linkage and Impact Fees

Another tool used to fund the construction of affordable housing is to charge fees on nonresidential development, such as new office buildings. This form of revenue generation is called a linkage fee. It is so named because it acknowledges the obvious link between commercial development and the subsequent and related demand for housing. By charging a linkage fee, the nonresidential developer supports, in part, the provision of affordable housing that is needed as a result of the new development. Without such linkage fees, the development of new commercial projects could lead to a greater population, thus increasing demand for housing, and producing higher housing costs. Without linkage fees, there is no mechanism to create additional affordable housing that may be needed as a result of new commercial development. Linkage fees are often tied to the permitting process and a project may not be approved until there is a contract that stipulates how the affordable housing conditions will be met. Depending on program design, developers may need to build new affordable housing units as part of their nonresidential development or pay into a local trust fund that is used for the construction of affordable housing or related activities. Like other regulatory programs, there are many factors that need to be considered in program design such as the geographic footprint in which linkage fees will apply, whether the program is voluntary or mandatory, and the acceptable uses for the funds that are raised.

A second type of fee charged to developers are one-time impact fees that cover the costs of infrastructure and services that are required as a result of new development. For example, these fees may be used directly to provide water, sewer, expand roads, or to fund new infrastructure of essential service such as police and fire stations, emergency services, or schools. Impact fees can support the development of affordable housing if that need is identified early during the capital facilities planning process. Lowering infrastructure costs decreases the financial burden for developers of affordable housing, as reduced infrastructure expenses can make the project financially feasible. The size of the fee is calculated using a formula based on a range of factors, including the project type, size and cost of the new development, demand generated by the new development, and the current level of interest rates.

Community Benefit Agreements

Community benefit agreements (CBAs) are an empowerment tool for residents to provide input throughout the development process for large commercial or residential projects. CBAs exist because there is a long history of existing residents being ignored during the development of new housing. As a result, new residents of the project are the primary beneficiaries, while incumbent residents are ignored, or even displaced. A CBA is a legally enforceable contract between existing community residents and the developer that outlines how the benefits of the new development will support the local community. Successful negotiations result in the developer gaining community support, which increases efficiency during the approval process and leads to a better development for the community. Affordable housing is one component that can be covered in a CBA. For example, when Facebook decided to do a major expansion of their East Palo Alto office, they entered a CBA with several community groups, creating a Catalyst Housing Fund. Facebook contributed an initial $20 million to the fund and the coalition was able to leverage about $60 million in funding. The money that was raised was dedicated to addressing the affordable housing crisis in the area by both developing and preserving affordable housing near the Facebook office complex (Sethi, 2020).

Regulations to Stabilize Rent

As rents continue to rise in many communities throughout the U.S., there are increasing calls for rent regulation. One of the challenges with this debate are the different types of rental regulations, including rent control and other forms of rent stabilization. One approach is affordability periods for affordable housing programs, which require projects to maintain prescribed levels of affordability for a given period of time as required in the LIHTC program. A second type includes a range of rental price mechanisms, such as rent control and rent stabilization. This type seeks to control the rents charged in the private rental market. Therefore, the first type seeks to control the cost of housing that is constructed with the benefit of public support or subsidy. The second form seeks to control rents being charged in the private market. Not surprisingly, the second category receives far more attention and is subject to rigorous debate on the wisdom and efficacy of these approaches.

Affordability Periods

Many rental housing programs have affordability provisions that require housing units built with governmental assistance to remain affordable for a set amount of time. Under these provisions, units constructed with the support of these funds must remain affordable to households who are at a certain income level for the control period, before the units can convert to market rate pricing. This provision guarantees that the funds used to create affordable housing benefit both initial occupants and future households. The Low-Income Housing Tax Credit (LIHTC) program, which is federally administered, is a prime example of how affordability periods affect the cost of housing constructed. The LIHTC program requires units to remain affordable for at least 30 years but many projects have longer affordability restrictions since the longer affordability window increases the chance a project will be awarded tax credits during the application process. State and local rental housing programs often incorporate control periods as well.

Although affordability periods preserve long-term affordability, developments subject to price controls may encounter capital improvement challenges. The lower rents charged in these projects may restrict the amount of funds available to address major capital needs throughout the life of a building. As a result, multi-family properties with no market rate units often need to raise additional funds for major improvements. The 4% tax credit through the LIHTC program is one funding source frequently used to support a substantial renovation of an affordable project. Another option for preserving the affordability of affordable housing developments is lifecycle underwriting. Typically, the costs for a multi-family development are based on financial viability for 10–20 years, whereas lifecycle underwriting considers the costs of maintaining the property for up to 50 years. Instead of facing a series of costly, one-time capital costs over the life of a building, 50-year lifecycle underwriting can make the project financially viable through the entire control period at a relatively modest cost (Brennan et al., 2013).

One of the key debates about control periods is what happens at the expiration of the period. As of 2023, a major concern for the affordable housing industry is the expiration of affordability requirements for hundreds of thousands of housing units that were constructed with support from LIHTC. According to research from the National Low Income Housing Coalition, between 2020 and 2030, nearly 500,000 units of affordable housing could be lost as affordability restrictions lapse (Aurand et al., 2018). In a nation that faces such significant shortages of affordable housing, losing hundreds of thousands of units would have significant negative consequences. As a result, this issue has received considerable attention from policymakers and advocates.

Rental Price Mechanisms

There is growing interest in rent regulation as rental costs continue to rise throughout the country. In the absence of other interventions to bring down the cost of rental housing, advocates and tenants increasingly push for more aggressive regulations to control rents. There are two primary forms of rent regulation: rent control and rent stabilization. Rent control, also referred to as first-generation rent control, places annual limits on how much a unit can be leased or renewed for and it nominally freezes rents by setting a rent ceiling. This was popular in New York during World War II, and persists to this day. Rent stabilization – also known as second-generation rent control – is a less aggressive approach. Stabilization seeks to keep the cost of rental housing aligned with inflation. Under this approach, rents can increase but the level of increase is limited to prevent rents from rising too rapidly. For example, the recent rent stabilization legislation in the state of Oregon limits annual rent increases to 7% plus inflation subject to a cap of 10%. This approach prevents rent gouging, but it does permit steady increases in rents, which can still lead to unaffordable housing over time.

While New York and San Francisco represent legacy models of rent control, there has been increased regulatory activity toward rent stabilization. As mentioned above, the State of Oregon passed a rent stabilization bill that covers the entire state, and California has passed a similar law. Many local jurisdictions have also adopted rent stabilization laws. It's important to underscore that even within the broader category of rent stabilization policies, there is significant policy variation. Key decisions that jurisdictions face when establishing a rent stabilization policy are: what units will be subject to the policy, how rents will be established under the policy, and finally whether there will be vacancy control (Office of Policy Development and Research, 2022). In determining which units will be subject to rental regulation, a key decision is whether to include or exclude new construction. Because of concerns that new development will be stifled by rent regulations, some jurisdictions exempt new construction from rent stabilization policies. Some locations exempt new construction for periods between five (Takoma Park, MD) and thirty years (Newark, NJ), while others provide no such exclusions for new construction (St. Paul, MN). In terms of price controls, rent regulation can range from relatively strict – St. Paul limits annual increases to 3%, while Oregon established a yearly level of 7% plus inflation. These different standards have obviously different implications for affordability and may – although the evidence is scarce – have different effects on new construction and development. Some states, like Arizona, Texas and Idaho, have gone in a completely different direction and have implemented preemptive measures against municipalities adopting rent control laws.

Finally, the vacancy decontrol mechanism can have a significant impact. Because annual rent increases are tied to a specific tenant, a new tenant – in the absence of other regulations – may provide a landlord with the opportunity to meaningfully increase the rents for a unit while still complying with existing rent regulations. The departure of a tenant would allow a landlord – absent vacancy control – to step up rents to prevailing market levels. As a result, there may be an incentive for landlords to consistently push rents up as much as possible and potentially force tenants to leave.[4] When the tenant leaves, the landlords may create a new base rent for the next tenant. To combat this practice, some rent regulation laws have established a vacancy decontrol mechanism that prevents landlords from increasing rents meaningfully when there is a change in tenant. The goal is to limit rent growth over longer periods of time, not just during the duration of one's tenancy in a particular unit. Vacancy decontrol mechanisms regulate the unit when it is vacant by providing guidelines as to how high rents can be set based on the previous rent of the unit.

In some jurisdictions, like New Jersey, rent control includes hardship exemptions that allow landlords to raise rents above what would otherwise be allowed (Seymour et al., 2024). These provisions were established in response to concerns that rent regulations might lead to degraded housing quality as a result of constrained rents that prevent adequate maintenance. Other provisions may exempt units that have been recently constructed or units that already receive some form of government assistance.

Not surprisingly, rent regulation has become highly politicized. There are strong advocates on both sides of the issue. Landlord groups, such as the National Apartment Association, aggressively lobby to prevent the expansion of rent control laws. On the other side, tenant organizations, such as Renters Rising, apply significant pressure on politicians to adopt more stringent rental regulations. In response to this back and forth, some states have prohibited local rent regulation ordinances altogether (Rajasekaran et al., 2019). Texas is one such state that has preempted local decision-making on matters related to rent regulation. Even if policymakers in a city in Texas had support to pass a rent stabilization law, it is prohibited because of the state action. The issue of rent regulation is one of the most hotly contested topics in the field of affordable housing. As long as housing costs remain unaffordable for so many people in the U.S., calls for greater rent control are likely to persist, and perhaps grow.

The arguments made by both sides in this debate are not surprising. Proponents of rent regulations argue that such laws can be used to address market failures that have produced housing that is unaffordable. Enacting these laws will allow people to keep their housing and prevent displacement, so goes the argument (Demsas, 2021; Horowitz & Starling, 2022). Opponents suggest that rent regulations are counterproductive because they are a source of inefficiency that harms landlords, restricts supply, and makes existing housing more expensive. The argument continues by suggesting that new capital investments in rental units may decrease as a consequence further restricting the supply of rental housing (Demsas, 2021; Horowitz & Starling, 2022). What is complicated about this debate is that there is empirical evidence to support both sides of the argument. The balance of the evidence suggests that the claims of both sides may, in part, be true (Demsas, 2021; Horowitz & Starling, 2022). A recent study conducted in San Francisco found rent control does prevent displacement of existing residents and can provide short-term rent relief, but it can also restrict housing supply as landlords shift units to owner-occupancy (i.e., condo conversion) which may increase market rents in the long run (Diamond et al., 2019). One remedy to address the adverse rental supply impact is to restrict condominium conversions as part of a rental regulation program.

Over the last several decades researchers have worked to provide empirical evidence on the impact of rental regulation but have been faced many challenges. Since each program is designed differently and housing markets within and across metropolitan areas vary, it is difficult to generalize findings. Researchers also have difficulty capturing outcomes associated directly with rent regulation policy itself since the programs are often designed with other policies and factors such as changes in demand and demographics impact the housing market. Most studies focus on New York City since it is one of the only places in the United States that has had both rent control and rent stabilization policies for decades. Considering these limitations, key findings from the literature reveal that even though in the short-term current tenants may benefit from reduced rents and displacement, there are many costs associated with rent regulations including:

1. benefits are poorly targeted with higher income households benefiting from policies intended to reach low-income households;
2. tenants in rent-controlled units tend to remain in place even if the unit no longer meets their needs which leads to inefficient allocation and over/under consumption of housing – many times because there are no other housing options for these tenants;
3. maintenance is often deferred in rent-controlled buildings leading to decay in the rental housing stock, which may suggest a need for broader code enforcement for stabilized properties to ensure landlord compliance;
4. there may be a reduction in rental housing supply from stricter forms of rent control which can lead to higher rents; and
5. rental regulations require resources for implementation and administration.

(Diamond, 2018; Sayin, 2020; Sturtevant, 2018)

In sum, more recent rent stabilization bills seek to deliver the benefits of rent control without the negative supply impacts. Examples include the 2019 rent stabilization bill in Oregon that caps rent increases at 10% but exempts all units constructed within the last 15 years. Such efforts seek to provide rent stability without stifling the development of new housing.

Conclusion

This chapter describes a range of regulatory approaches at the federal, state, and local level to limit discriminatory practices, reduce residential segregation, expand housing choice, and promote increased housing affordability. At the federal level, the Fair Housing Act and the Community Reinvestment Act have been instrumental in reducing discrimination in the housing market. At the state and local level, regulatory mechanisms such as zoning reform, inclusionary housing, and incentives to encourage infill housing are used to produce or promote affordable housing through residential densification. States and local jurisdictions also use a range of tools to facilitate and finance affordable housing through development provisions like expedited permitting, linkage and impact fees, and community benefits agreements. Finally, states and localities are actively engaged in either promoting or preventing the passage of rent regulations. The complexities associated with understanding the effects of rent regulation make this a hotly-contested corner of the affordable housing industry. In sum, the mechanisms described in this chapter are a toolkit from which policymakers may select to create or promote the creation of more affordable housing in their communities.

Notes

1 https://therealdeal.com/sanfrancisco/2023/07/26/sf-cuts-affordable-requirements-and-fees-to-spur-housing-growth/
2 Opponents might argue that more units of housing and greater supply of housing could have been created using different policy tools.
3 See the full text of Executive Directive 1 here: https://planning.lacity.gov/odocument/e27072a3–7da3–4120-b94b-1e0874642f03/ED_1_-_Expedition_of_Permits_and_Clearances_for_Temporary_Shelters_and_Affordable_Housing_Types_Revised.pdf
4 There is a relationship between rent regulation and eviction regulation. For example, good cause eviction laws require a landlord to provide "good cause" for terminating tenancy. These protections also require advance notice from a landlord before ending a lease. Therefore, vacancy decontrol paired with good cause eviction protections may provide a more robust rent regulation.

References

Aurand, A., Emmanuel, D., Stater, K., & McElwain, K. (2018). *Balancing priorities: Preservation and neighborhood opportunity in the low-income housing tax credit program beyond year 30*. Washington, DC: National Low Income Housing Coalition and the Public and Affordable Housing Research Corporation.

Badger, E., & Bui, Q. (2019). Cities start to question an american ideal: A house with a yard on every lot – townhomes, duplexes and apartments are effectively banned in many neighborhoods. Now some communities regret it. *The New York Times*. Retrieved from www.nytimes.com/interactive/2019/06/18/upshot/cities-across-america-question-single-family-zoning.html?mtrref=undefined&gwh=FE3E0E6DEA0C197CA95B7B80FEC540DB&gwt=pay&assetType=PAYWALL

Brennan, M., Deora, A., Handelman, E., Heegaard, A., Lee, A., Lubell, J., & Wilkins, C. (2013). *Lifecycle underwriting: Potential policy and practical implications*. Center for Housing Policy & National Housing Conference. Retrieved from http://docs.wixstatic.com/ugd/19cfbe_891b4788e2e64d0cb71a75940a101f2f.pdf

Britschgi, C. (2022). *Eliminating single-family zoning isn't the reason minneapolis is a YIMBY success story*. Retrieved from https://reason.com/2022/05/11/eliminating-single-family-zoning-isnt-the-reason-minneapolis-is-a-yimby-success-story/#:~:text=Minneapolis%20appears%20to%20be%20a,do%20with%20this%20success%2C%20however.

Christopher, B. (2024). Los Angeles' one weird trick to build affordable housing at no public cost. *Cal Matters*, February 7, 2024. Retrieved from https://calmatters.org/housing/2024/02/affordable-housing-los-angeles/

Demsas, J. (2021). *I changed my mind on rent control. Rent control won't fix the housing crisis. It's still a good idea*. Retrieved from www.vox.com/22789296/housing-crisis-rent-relief-control-supply

Diamond, R. (2018). *What does economic evidence tell us about the effects of rent control?* Brookings. Retrieved from www.brookings.edu/articles/what-does-economic-evidence-tell-us-about-the-effects-of-rent-control/

Diamond, R., McQuade, T., & Qian, F. (2019). The effects of rent control expansion on tenants, landlords, and inequality: Evidence from San Francisco. *American Economic Review, 109*(9), 3365–3394.

Einstein, K., L., Glick, D. M., & Palmer, M. (2019). *Neighborhood defenders: Participatory politics and America's housing crisis*. Cambridge: Cambridge University Press.

Friedman, S., & Squires, G. D. (2005). Does the Community Reinvestment Act help minorities access traditionally inaccessible neighborhoods? *Social Problems, 52*(2), 209–231.

Goodman, L., Zhu, L., Seidman, E., Ratcliffe, J., & Zhu, J. (2022). *Should the community reinvestment act consider race?* Urban Institute. Retrieved from www.urban.org/sites/default/files/2022–04/should-the-community-reinvestment-act-consider-race_1.pdf

Gyourko, J., & Molloy, R. (2015). Regulation and housing supply. *Handbook of Regional and Urban Economics, 5*, 1289–1337.

Horowitz, B., & Starling, L. (2022). *An overview of rent stabilization from national housing experts*. Retrieved from www.minneapolisfed.org/article/2022/an-overview-of-rent-stabilization-from-national-housing-experts

Lee, L. (2022). *How eliminating single-family-only zoning will impact builders and developers.* Retrieved from www.builderonline.com/land/planning/how-eliminating-single-family-only-zoning-will-impact-builders-and-developers_o

Mukhija, V., Das, A., Regus, L., & Tsay, S. S. (2015). The tradeoffs of inclusionary zoning: What do we know and what do we need to know? *Planning Practice and Research, 30*(2), 222–235.

National Low Income Housing Coalition (NLIHC). (2020). *Trump administration eliminates affirmatively furthering fair housing rule, NLIHC and other advocates condemn action, rhetoric.* Retrieved from https://nlihc.org/resource/trump-administration-eliminates-affirmatively-furthering-fair-housing-rule-nlihc-and-other

Office of Policy Development & Research. (2022). *Options and tradeoffs: Rent stabilization policies.* Washington, DC: U.S. Department of Housing and Urban Development.

O'Regan, K. M. (2018). The Fair Housing Act today: Current context and challenges at 50. *Housing Policy Debate, 29*(5), 704–713.

Phillips, S. (2024). *Modeling inclusionary zoning's impact on housing production in Los Angeles: Trade-offs and policy implications.* Terner Center for Housing Innovation, University of California, Berkeley. Retrieved from https://ternercenter.berkeley.edu/wp-content/uploads/2024/04/Inclusionary-Zoning-Paper-April-2024-Final.pdf

DC Policy Center. *Appraising the district's rentals.* Retrieved from www.dcpolicycenter.org/publications/appraising-districts-rentals/

Rajasekaran, P., Treskon, M., & Greene, S. (2019). *Rent control: What does the research tell us about the effectiveness of local action?* Urban Institute. Retrieved from www.urban.org/sites/default/files/publication/99646/rent_control._what_does_the_research_tell_us_about_the_effectiveness_of_local_action_1.pdf

Reid, C. (2019). *Quantitative performance metrics for CRA. How much "reinvestment" is enough?* Penn Institute for Urban Research. Retrieved from https://penniur.upenn.edu/uploads/media/Quantitative_Performance.pdf

Rothstein, R. (2017). *Color of law: A forgotten history of how our government segregated America.* New York: Liveright Publishing Corporation, a Division of W. W. Norton & Company.

Sayin, Y. (2020). *Appraising the district's rentals. Appendix II – review of literature on the impact of rent control on housing quality and quantity, displacement, and inclusion.* DC Policy Center.

Sethi, M. (2020). *Facebook's catalyst fund creates hundreds of affordable homes near menlo park headquarters.* Retrieved from https://about.fb.com/news/2020/08/catalyst-fund-creates-affordable-homes/

Seymour, E., Payne, W., Newman, K., Deitz, S., & Nolan, L. (2024). *Rent control in New Jersey. What is it? Where is it? How does it work?* Ralph W. Voorhees Center for Civic Engagement Edward J. Bloustein School of Planning and Public Policy Rutgers University.

Sturtevant, L. (2018). *The impacts of rent control: A research review and synthesis.* Washington, DC: National Multifamily Housing Coalition.

Wang, R., & Balachandran, S. (2021). *Inclusionary housing in the United State. Prevalence, practice, and production in local jurisdictions as of 2019.* Washington, DC: Grounded Solutions Network. Retrieved from https://groundedsolutions.org/tools-for-success/resource-library/inclusionary-housing-united-states#:~:text=At%20the%20end%20of%202019,mandatory%20programs%20as%20voluntary%20programs.

Part IV

Case Studies

Chapter 12

Chicago, IL

Overview

Behind only New York and Los Angeles, Chicago is the third largest city in the United States. On the shores of Lake Michigan, Chicago is known for its natural beauty, world class architecture, and a thriving downtown with amenities for all tastes, including the arts, professional sports, and a highly regarded restaurant scene. Chicago has served as the economic engine of the Midwest for much of the last 150 years. The city was the rail hub of the nation, agriculture was grown in the region and transported out of the city by ship, train, or truck; the Chicago Board of Trade provided markets to buy, sell, and hedge exposure to commodities; Lasalle Street became known as the Midwestern Wall Street; steel was manufactured and shipped via the Great Lakes; and Sears, Roebuck, & Co. became the dominant retailer in the nation for nearly 100 years. While not solely dependent upon one industry – as Detroit was with automobile manufacturing – the Chicago economy was still heavily reliant on manufacturing. As global supply chains became a reality, fewer goods were being manufactured in the United States. Cities such as Chicago, Detroit, Cleveland, and Pittsburgh all felt the consequences of this global economic shift. As a result, Chicago faced a slow and steady decline in population from 1950 on. This economic stagnation put Chicago on a different trajectory than the other major cities in the United States that continued to grow and prosper throughout the second half of the 20th century.

Another major factor in the development of Chicago was the Great Migration, in which households of color moved to northern cities. Between 1910 and 1970, approximately six million Black people moved from the southern United States to escape rampant racial segregation that was common in the Jim Crow south. This mass movement of people radically reshaped many northern cities, including Chicago. These migrants from southern states formed the basis of the Black community in Chicago. While escaping the conspicuous prejudice in the south, Black households did not move into equitable and integrated communities. Rather, many northern cities became some of the most segregated cities in the nation, and Chicago was at the top of this list. Black households have tended to live in the southern and western neighborhoods of Chicago, while white households have disproportionately lived in wealthier neighborhoods in the northern part of the city. These racial dynamics play an important role as we consider the housing history of Chicago.

Chicago played an important role in the trajectory of affordable housing policy in the United States since the end of World War II. Home to a large, and notorious, stock of public housing, Chicago became synonymous with urban failure and decline, and housing played a critical role in this narrative. For a nation unfamiliar with the details of affordable housing policy, mental images about public housing projects were framed by the experiences in Chicago

DOI: 10.1201/9781003356585-16

and media depictions of the events. Even in popular culture, Chicago and "the projects" were linked. In the 1970s, CBS aired a popular sitcom called, Good Times, which followed family members living in a public housing project in Chicago – based on the infamous Cabrini-Green Homes.

Chicago also played an influential role in a key transition in affordable housing policy in the United States. In 1966, Dorothy Gautreax, a resident of public housing in Chicago, filed a class action lawsuit against the Chicago Housing Authority (CHA) and U.S. Department of Housing and Urban Development (HUD). The lawsuit alleged that CHA had contributed to racial segregation in the public housing program in Chicago. It also claimed that HUD had contributed to this segregation in the way it funded public housing programs. In response, the Gautreaux Program was established which provided mobility options for Black families living in public housing; many of these families moved to affluent, white communities. The program laid the foundation for future ongoing policy experiments designed to provide choice and opportunity to low-income (and frequently Black) households. The Housing Choice Voucher program and the Moving to Opportunity experiment are two prime examples of how choice and mobility have been incorporated into federal housing policy.

The face of affordable housing in Chicago has changed considerably in the last thirty years. Many of the public housing towers that were an integral part of the affordable housing landscape in Chicago – in the second half of the last century – are no longer standing. For example, the final tower of what once was the Cabrini-Green housing project was taken down in 2011. In its place, lower density, mixed income developments that were desired by city leaders were constructed, but only about 11% of public housing residents returned to the new developments. The Chicago Housing Authority's Plan for Transformation aimed to demolish approximately 22,000 public housing units, renovate 17,000 units, and construct another 7,700 mixed-income developments.[1] The promise of this transformation was not fully realized which continues the legacy of hardship faced by households served by the Chicago Housing Authority.

Demographic Overview

As of 2022, Chicago was home to a population of approximately 2.7 million people across 1.1 million households. Chicago is also one of the ten most dense cities in the United States. But interestingly, the story of Chicago is not one of rapid population growth – at least in recent history. In the 1890 census, Chicago was the fifth largest city in the world. And by 1950, Chicago had a population that exceeded 3.6 million people. Since 1950, there has been a steady decline in the city's population – the city is now home to roughly 900,000 fewer people than lived there in the middle of the last century. The vast majority of this population decline occurred during the forty years between 1950 and 1990. Since 1990, the population of Chicago has stayed relatively constant. These demographic patterns have important implications for housing and affordability.

The decline of Chicago's population is part of the broader population shift away from the industrial Midwest. Chicago's declines are not nearly as pronounced as what was experienced by its Rust Belt siblings, Detroit and Cleveland, but similar demographic patterns are at play. Further evidence of this broader economic decline is the relatively low household incomes in Chicago. As of 2022, the median household income in Chicago was $70,386 which was less than the U.S. median of $74,755. Consistent with the income statistics, Chicago's poverty rate was 17.2% in 2022, far higher than the national rate of 12.6%. In sum, Chicago – despite its beauty and vibrancy – is a city with lower incomes, higher poverty, and stagnant population growth.

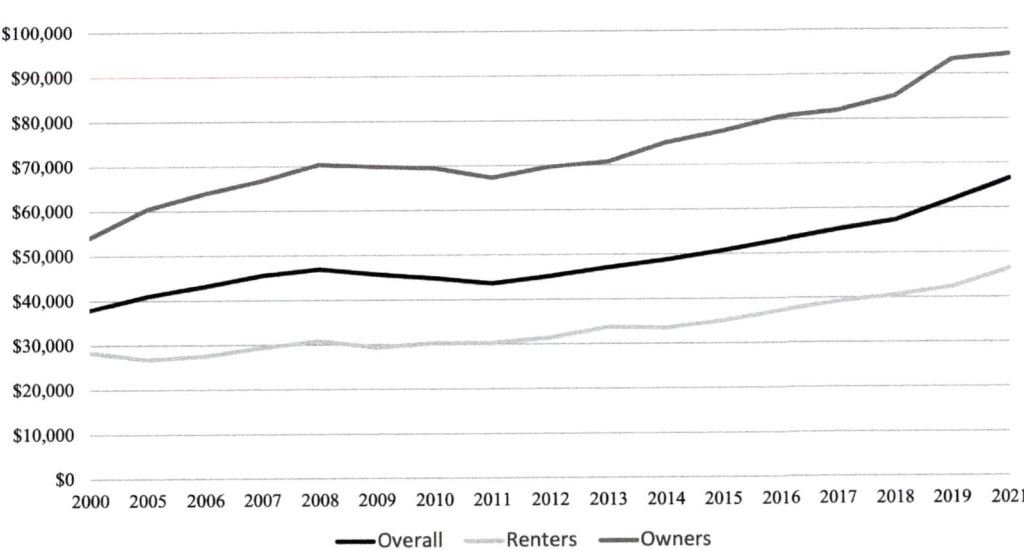

Figure 12.1 Median Household Income in Chicago, by Housing Tenure

(Source: American Community Survey 1-Year Estimates, Decennial Census)

Like many other communities in the United States, there is a pronounced difference in the household incomes of homeowners and renters. Figure 12.1 highlights the trend in household income broken down by housing tenure in Chicago. Over the last 20 years, there has been a persistent gap between renters and owners and, by 2021, homeowners had nearly double the household income of renters who earned less than $50,000 in annual income.

Housing Stock and Market Overview

According to the 2022 American Community Survey, Chicago is home to 1,159,424 households. But the housing stock in the city exceeds the number of households, which is not unusual. The total housing stock in Chicago is 1,262,463 units. This suggests that there are 103,039 total vacant housing units in the city (Table 12.1). These vacancy numbers can be source of confusion for many observers, so it is important to dissect these numbers. The current vacancy rate for homeownership is only 0.8% and the rental market vacancy rate is 4.2%. Yet, the percent of vacant housing units in the City of Chicago is 9.7%.

Table 12.1 breaks down the 103,039 total vacant units in the city:

Table 12.1 Vacant units in the City of Chicago in 2022

For rent	27,771
Rented, not occupied	3,812
For sale only	5,113
Sold, not occupied	1,754
For season, recreational, or occasional use	11,825
Other vacant	52,678
For migrants	86
Total vacant units	103,039

Table 12.2 Composition of Housing Stock in City of Chicago, 2022

Units in structure		
1-unit detached	320,666	25.4%
1-unit attached	47,974	3.8%
2–9 units	487,311	38.6%
10 or more units	402,726	31.9%
Mobile homes	3,787	0.3%
Total housing units	1,262,463	

As Table 12.1 highlights, there are only 27,771 units available for rent at the time of this measure. We arrive at the rental market vacancy rate of 4.2%, by dividing the number of units available for rent by the sum of the total occupied rental units plus those available for rent. As one can see, the number of vacancies that are undefined and those that are only used for part of the year are large contributors to the level of vacancy in the city. There are open questions about what constitutes the "Other vacant" category. It could include short-term rentals or investment properties that might be sitting vacant. This is an area where researchers need to focus and the U.S. Census needs to provide greater information.

Chicago stands in contrast to many other locations in the country with respect to the composition of its housing stock. Unlike many other jurisdictions in the United States where detached single-family homes are the dominant housing type, that is not the case in Chicago. This category of housing only represents just over a quarter of all housing units in the city. Instead, small apartment buildings are the most common type of housing structure in the Chicago. Table 12.2 breaks down the city's housing stock by type.

In terms of household structure and housing tenure, Chicago is a city of majority renters. As shown in Figure 12.2 below, during the housing boom of the late 2000s, homeownership in Chicago nearly reached 50%, but since the bursting of the housing bubble, levels of homeownership

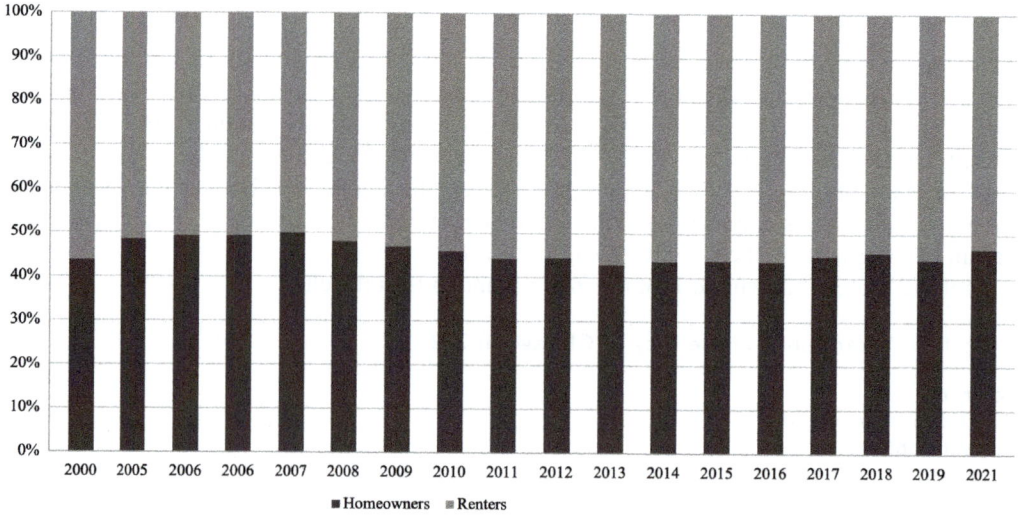

Figure 12.2 Housing Tenure, City of Chicago, 2000–2021

(Source: American Community Survey 1-Year Estimates, Decennial Census)

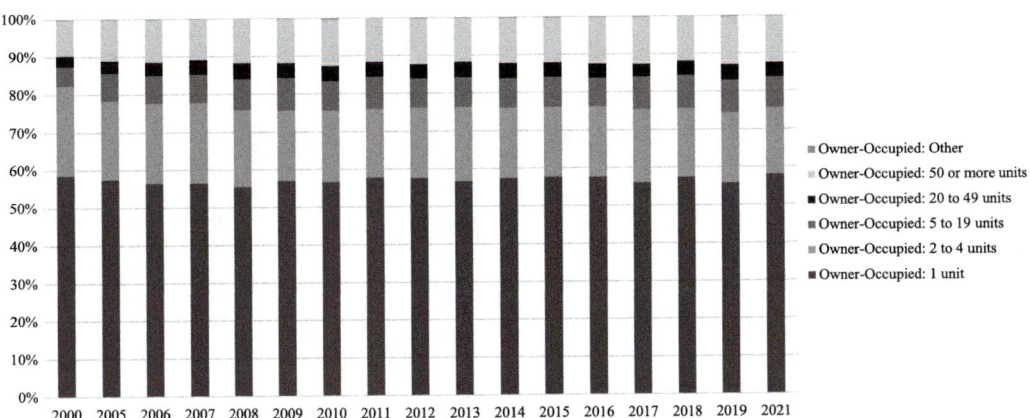

Figure 12.3 Owner-Occupied Housing by Number of Units, 2000–2021
(Source: American Community Survey 1-Year Estimates, Decennial Census)

have settled at roughly 45% of households. Therefore, policies and market dynamics that affect rental housing are very important given the disproportionate reliance on rental housing by Chicago households.

Like most places in the United States, a majority of homeowners reside in one unit structures. Detached, single-family housing remains the dominant form of housing in the nation. But given the density of housing in Chicago, a greater percentage of people who own their housing reside in multi-unit dwellings. Owner-occupied single unit structures account for less than 60% of all owner-occupied housing. According to research from the Institute for Housing Studies at DePaul, there has been an increase in the loss of 2- to 4-unit buildings in all neighborhoods in Chicago. Particularly in higher cost neighborhoods, these 2- to 4-unit structures are being replaced by single-family homes which negatively impacts the supply of housing and reduces affordable housing options that are common in these small, multi-unit structures (Institute for Housing Studies at DePaul University, 2021). Figure 12.3 highlights the breakdown of housing type among homeowners in Chicago.

Consistent with intuition, the percent of renters residing in multi-family housing is much higher than for homeowners. As highlighted in Figure 12.4, the majority of renter households reside in buildings with 2 to 19 units. There has been an increase in the share of units in large (50 or more units) buildings. In 2000, about 20% of renter households resided in large buildings and the share increased to about 25% by 2021. In sum, rental housing in Chicago has become increasingly dense over the last two decades.

One of the hallmarks of cities with declining or stagnating population is greater affordability. The key reason for this relationship is the durability of housing. Because housing lasts a long time (hundreds of years in some cases), when people leave a city, the housing remains. If you have a balanced housing market (sufficient supply for the existing demand for housing), and a city loses 5% of its population, the city will have excess housing supply given the reduction in population. This dynamic helps to explain the extreme affordability in Rust Belt cities such as Detroit and Cleveland. Chicago has also been a beneficiary of this relationship, but to a lesser extent. The population decline in Chicago – and resulting housing affordability – has been more muted than in other post-industrial cities.

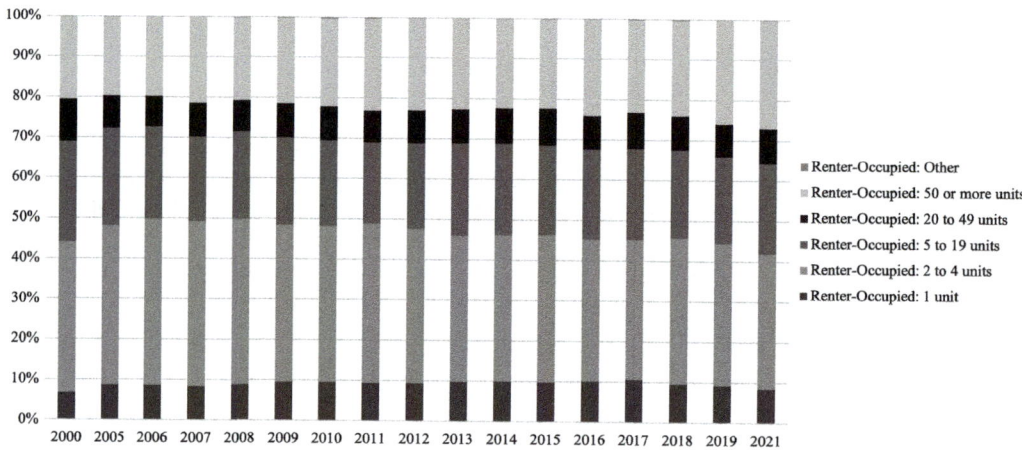

Figure 12.4 Renter-Occupied Housing by Number of Units, 2000–2021
(Source: American Community Survey 1-Year Estimates, Decennial Census)

Given these population dynamics, theory and intuition suggest that Chicago will have lower housing costs than other large U.S. cities, most of which have continued to grow. The data support this conclusion. In 2022, median gross rents in Chicago were $1,324 a month compared to a national average of $1,300. In New York and Los Angeles, median gross rents were $1,688 and $1,788 respectively. Among the three largest cities in the United States, Chicago is far more affordable. In San Francisco, the equivalent rent is $2,308, almost a thousand dollars more per month than in Chicago.

As discussed earlier in the book, lower rents do not necessarily imply affordable rental housing. The lower rents enjoyed by residents in Chicago are accompanied by lower household incomes. As a result, the housing cost burden faced by Chicago households is not radically different than the burdens faced by households in many other communities throughout the nation. In 2022, nearly 52% of all Chicago renter households were housing cost burdened (they paid more than 30% of their income toward housing and utility costs). In San Francisco, about 38% of renter households are burdened. Among the largest cities, Chicago is comparable – 59% of Los Angeles renters are cost burdened, while that figure is 52% in New York.

A focus on homeownership highlights the pronounced differences between Chicago and other large cities in the United States. According to the American Community Survey, the median value of owner-occupied housing units in Chicago in 2022 was $313,300. This was far lower than the other large cities that we've highlighted in this section. The comparable figure was $724,400 in New York, $903,700 in Los Angeles, and a staggering $1,343,700 in San Francisco.

According to research from The Institute for Housing Studies at DePaul University, rental affordability in Chicago has deteriorated over the last decade. The Institute calculates a gap in affordable rental housing by comparing the number of low-income households to the number of rental units that are affordable to this group of renters. In 2021 (the most recent year for which data is available), this gap was 119,435 units, the highest in a decade. The authors indicate that the primary driver of the increase in the gap was the decline in the number of affordable rental units in Chicago. The numbers from researchers at DePaul University highlight the troubling trend of decreased housing affordability in our nation's third largest city (Institute for Housing Studies, 2023).

To put the gap in affordable housing in perspective, the total stock of housing in Chicago is about 1.26 million units. Therefore, the 119,435 unit gap in affordable housing is nearly 10% of the total housing stock in Chicago. Therefore, significant scale is needed – rapidly – in order to meet the need. The scale of construction would need to significantly increase relative to the rate of housing production growth over the last decade. From 2012 to 2022, the housing stock in Chicago increased by just over 73,000 housing units, an increase of 6.2%. This is solid growth, but in light of the well-documented gap in affordable housing, this pace of growth won't be sufficient. Greater focus, attention, and resources are necessary to close this gap. Some of this gap could be filled by existing housing inventory by providing households additional rental assistance such that they can access more expensive housing (that might be available) in the private rental market.

Chicago, therefore, is a city of contrasts. For low-income households, housing remains very unaffordable despite lower absolute rent levels. This relationship is driven by relatively high levels of poverty and low household income levels combined with an inadequate supply of affordable housing. But for higher income households, Chicago provides excellent housing access. Rents are relatively affordable and the price to purchase a home is significantly cheaper than in other large cities in the United States. Chicago highlights the clear message that lower rents are not a cure for all housing affordability challenges. The persistent problem of low incomes continues to complicate efforts to ensure housing access and affordability.

Federal Housing Supports

Against the backdrop of Chicago's demographics, housing stock, and housing market, we next explore how support from the federal government helps to promote housing access in Chicago. This analysis is difficult given the variety of data sources involved, incomplete data, and the various subsidy programs that exist at multiple levels of government. Therefore, this analysis is incomplete and should only be considered a broad summary of subsidized housing in the City of Chicago.

According to HUD's Picture of Subsidized Households database, in Chicago there were roughly 105,000 housing units that that received some type of federal housing support in 2023. This represented over 8% of all housing units in the city. By comparison, the equivalent figure for the nation as a whole was roughly 3.6%. Therefore, Chicago received disproportionate support from federal housing programs than did the rest of the country. One explanation for this variation is due to the legacy of housing programs. When established, federal housing programs provided benefits to large cities which, at the time were located in the Northeast, mid-Atlantic, and industrial Midwest. Therefore, as the population of the United States dispersed to the south and west, the allocation of benefits didn't follow suit. As a result, we see gaps in the percentage of housing units subsidized by geography based on the history of these housing programs.

In addition to the federal programs outlined above, the LIHTC program has been used to create affordable housing units in Chicago. According the HUD LIHTC database, there are over 52,000 LIHTC units available to low-income households. These LIHTC units represent about 8% of the entire rental housing stock in the city. These units, therefore, play an important role in creating affordable housing options for low-income households in Chicago.

The significant presence of federal housing support combined with a robust supply of LIHTC units supports a significant stock of households that receive housing support. Because we know that there is overlap in housing subsidies – some people who live in LIHTC units also receive rental support – these subsidies are not mutually exclusive (Colburn et al., 2024). Therefore, if we assume that 30% of people who live in LIHTC units also receive housing support, that would imply that there are about 142,000 units of housing that receive federal housing support, which is over 12% of all housing units in Chicago. This is far higher than the national average of about 5% of all households that receive federal support (rental assistance or LIHTC). But even with the disproportionate benefits allocated to Chicago, the level of federal support is inadequate relative to the poverty rate of over 17%. There are still many households in poverty that do not receive federal housing assistance.

State and Local Housing Supports

For residents of Chicago, there are a range of state and local policies and programs that supplement the assistance provided by federally-funded programs. This is common in most jurisdictions as lower levels of government seek to address unmet needs in their communities. As described elsewhere in this book, these policies and programs are not necessarily independent of one another. Rather, assistance from different levels of government might be used by the same household or development.

In 2018, the City of Chicago produced *One Chicago: Housing Strategies for a Thriving City, Five Year Housing Plan, 2019–2023*. Developed by the city's Department of Planning and Development, the plan established a new Department of Housing in the City of Chicago. The city had previously had a housing department prior to 2008, but that function had subsequently been consolidated in the Department of Housing and Economic Development. Those functions have now been separated. This administrative change highlighted the importance of housing in the city and how the government can respond to the growing needs of its residents. This plan was the sixth city plan for housing that covered the preceding thirty years. According to *One Chicago*, the Department of Housing:

> will offer a seamless City partner to the many public, nonprofit and private organizations dedicated to housing in Chicago. It will continue to leverage the full gambit of City resources, including vacant land and buildings, TIF, tax-exempt bonds, the funding and additional affordable housing units through the Affordable Requirements Ordinance and other resources required to create a variety of housing options for all residents of Chicago. The Department will ensure the City is upholding its commitment to providing affordable housing that is accessible to people with disabilities.
>
> (City of Chicago, 2019)

The five-year plan outlined five ways to promote affordable housing in the city:

1 neighborhood focus, especially in locations facing gentrification or those facing disinvestment,
2 invest in affordable rental stock,
3 support housing for city's most vulnerable residents,
4 expand homeownership opportunities, and
5 promote housing innovation and collaboration.

As outlined in the city's plan, there are resources from a range of programs that can be activated to create and provide affordable housing. At the state level, Illinois has significant programs that provide resources for developers and households in Chicago. The Illinois Affordable Housing Trust Fund was established in 1989 and was funded with a $4.6 million commitment in 2012. The trust fund receives one half of the state's receipts from the Real Estate Transfer Tax. Proceeds from this fund can be used to "provide grants and loans for the acquisition, construction, development and rehabilitation, operation, insurance or retention of multi-family and single-family developments."

The state also established the Illinois Affordable Housing Tax Credit program which provides a $0.50 tax credit for every dollar invested in an affordable housing project by non-profit sponsored developments that target households with incomes below 60% AMI. Roughly 25% of all tax credits established by this program are allocated to the City of Chicago. The Illinois Affordable Housing Tax Credit program is not exclusively used with the federal LIHTC so it is not the same as many of the other state tax credit programs across the country that supplement federal tax credits. In fact, the State of Illinois is considering a state LIHTC program that would be more robust and would be similar to other state programs.

The Rental Housing Support Program (RHSP) was created by the State of Illinois in 2005 to provide rental assistance to Illinois' poorest households. It is a unit-based subsidy serving extremely low-income households with incomes at or below 30% area median income. The Illinois Housing Development Authority (IHDA) administers the program, contracting with Local Administering Agencies (LAAs) who manage the program in their communities. In Chicago, the program is administered by the Chicago Low Income Housing Trust Fund. The program was first funded by a $9 fee for each recorded real-estate document at the county-level. The state legislature increased the fee to $18 in 2023 to provide additional resources for the program.

Finally, the Illinois Housing Development Authority manages a low-income homebuying program. This initiative provides access to homeownership for households that may, under different circumstances, not have access to ownership. Through a range of supports, including down payment assistance and funding to pay for closing costs, the program provides a pathway to homeownership.

At the local level, the City of Chicago has also created a range of programs to supplement those that exist federally and at the state level. First, there is the Chicago Low-Income Housing Trust Fund (CLIHTF). The program provides a range of supports including subsidies for housing development, special programs for households with specific needs, and a rental subsidy program that has supported over 7,000 households. The CLIHTF is the largest state- and city-funded rental subsidy program in the nation. The trust fund is partially funded with corporate and in lieu fees through the Affordable Requirements Ordinance (described below). Additional support comes from the State of Illinois through the Rental Housing Support Program, which is funded by a recording fee collected by the county clerks and recorders and is administered by the IHDA. In addition to supporting new development and providing rental subsidies, the city has also created Chicago Neighborhood Rebuild Pilot Program to promote rehabilitation of vacant homes. While small, this program recognizes that Chicago has vacant legacy housing stock that is an asset for the community, but requires an investment to bring that housing back online. There is also a Chicago Community Housing Trust, which used to be known as the Chicago Community Land Trust (CCLT) that sells homes to eligible households at affordable prices. There are ongoing covenants attached to the housing to ensure long-term affordability. While none of these programs have significant scale, they still provide much needed support to households that are not served by the larger federal or state programs.

In addition to the homeownership and rental support programs outlined above, the city also has passed the Affordability Requirements Ordinance (ARO) which uses a number of regulatory provisions to promote greater housing affordability. The ordinance requires affordable units to be provided in certain developments that have ten or more units. The amount of required affordable units depends on the project type and ranges from 8% to 20%. In 2017, the ordinance was expanded to include gentrifying areas. As of 2020, the ARO has resulted in roughly 1,500 affordable housing units for household with low and very low incomes. In 2021, the state passed a major property tax incentive to further incentive affordability in ARO developments. Properties that have at least 20% affordable units receive a reduction in assessed value. Prior to the creation of the ARO, the city also established a density bonus program that provides additional building capacity (through greater density) to developers in exchange for providing on-site affordable housing units at that development. The developer has the option to pay into the housing trust fund in lieu of building onsite units. Chicago's extensive list of programs highlights how local jurisdictions can engage in creating, preserving, and subsidizing affordable housing.

Conclusion

Chicago has several advantages that set it apart from other large cities in the United States. First, it is one of the more affordable large cities. Second, it gets disproportionate support from the federal government. Third, the State of Illinois and City of Chicago have a robust set of programs that provide support on top of that provided by the federal government. But despite these inherent advantages, low-income households still face increasing difficulty securing housing that is affordable to them. This brief case study highlights that no city is without significant affordability challenges. The test for Chicago is to continue to provide sufficient resources to encourage the development and construction of affordable housing for low-income households that earn below 60% of area median income. According to DePaul University researchers, that gap in housing is nearly 120,000 housing units. Therefore, over the next decade, the City of Chicago will need to prioritize the development of additional housing, particularly affordable housing, if residents of all income levels can continue to live and thrive in the city by the lake.

Note

1 Public Housing in Chicago, www.illinoislegalaid.org/legal-information/public-housing-chicago

References

The Chicago Housing Authority. (2023). Retrieved from www.thecha.org/about

Affordable Requirements Ordinance (ARO). (2023). Retrieved from www.chicago.gov/content/city/en/depts/doh/provdrs/developers/svcs/aro.html

Bowly Jr., D. (1978). *The poorhouse: Subsidized housing in Chicago*, 2nd edition. Southern Illinois University Press.

City of Chicago. (2019). *One Chicago: Housing strategies for a thriving city, five year housing plan 2019 – 2023*. Department of Planning & Development. Retrieved from www.chicago.gov/content/dam/city/depts/dcd/Housing%20Programs/20733_37_5_Year_Plan_Report_final_WEB_C.pdf

Colburn, G., Acolin, A., & Walter, R. (2024). Subsidy overlap in federal housing policy. *Housing Policy Debate*.

Comprehensive Housing Planning Act, Pub. L. No. 094–0965. (2003). Retrieved from www.ihda.org/wp-content/uploads/2016/03/CompHsgPlanAct_PA94–965.pdf

Gautreaux v. Chicago Housing Authority, 296 F. Supp. 907 (N.D. Ill. 1969). Justia Law. Retrieved from https://law.justia.com/cases/federal/district-courts/FSupp/296/907/1982538/

Hills v. Gautreaux: 425 U.S. 284 (1976): Justia US Supreme Court Center (n.d.). Retrieved from https://supreme.justia.com/cases/federal/us/425/284/

Institute for Housing Studies at DePaul University. (2021). *Patterns of Lost 2 to 4 Unit Buildings in Chicago*. Retrieved from https://housingstudies.org/releases/patterns-lost-2-4-unit-buildings-chicago/

Institute for Housing Studies at DePaul University. (2023). *2023 State of Rental Housing in the City of Chicago*. Retrieved from www.housingstudies.org/releases/2023-state-rental-housing-city-chicago/

Smith II, P. H. (2012). *Racial democracy and the black metropolis: Housing policy in postwar Chicago*. University of Minnesota Press.

Chapter 13

San Antonio, TX

Background

For residents of coastal cities, San Antonio might appear as an affordable place to live, but many San Antonians would disagree. A notable proportion of the city's housing is older and in disrepair and often fails to meet basic standards of safe and sanitary conditions or is located in economically depressed areas. Additionally, the combination of low wages and escalating housing costs prevents many individuals from accessing newer, higher-quality housing options. As the population continues to grow rapidly, the limited housing supply further exacerbates the struggle for low- and moderate-income households to find suitable housing.

San Antonio is a fascinating case due to the significant historical events that have had a major impact on housing in the city. It became the largest city in Texas in 1860 and the growth of the railroad played a key role in the city's population growth and economic expansion. San Antonio, also known as the Alamo city, was the center of Spanish-Mexican conflicts, battles for Texas independence, and the U.S.-Mexican War. The city is home to a considerable immigrant and native population. During the early 1900s, there was a rise in Mexican immigration to San Antonio due to the Mexican Revolution. By the end of 1940, Mexicans made up over 40% of the city's population. Mexican immigrants were primarily offered low-skilled and temporary work opportunities. Due to the meager pay and general disinvestment, Hispanic communities located west of downtown San Antonio soon became known for having some of the poorest housing conditions in the United States (Zelman, 1983).

In the 1930s, the Texas Planning Board performed a study on low-income housing in San Antonio, identifying unsanitary living conditions in the Hispanic West side of the city. In the 1940s, 1,674 slum units were replaced with 2,554 public housing units under the Wagner-Steagall Act (Fairbanks, 2002). Alazan-Apache Courts, the first public housing development in San Antonio and among the first in the country, consisted of 1,180 single-family homes located on the city's west side (Zelman, 1983). The development still stands today but now includes a mix of multi-family and single-family units. Houston and Dallas, the two biggest cities in Texas, dismissed urban renewal initiatives in the 1950s due to concerns about socialist policies and the erosion of individualism. In contrast, San Antonio actively supported urban renewal initiatives to enhance the city's aesthetics, which unfortunately led to displacement of communities of color in areas that previously provided affordable housing options (Fairbanks, 2002).

The expansion of highway construction, along with urban renewal efforts, spurred suburban growth and decentralization. During the second half of the twentieth century, San Antonio expanded outward and produced a boom in the construction of single-family homes (Caine et al., 2017).

During this expansion, resources were directed toward the north and northwest areas of the city (which tended to be occupied by white households living in single-family homes), while the West and Southwest neighborhoods, predominantly Hispanic, and the Eastside, primarily African American, experienced sustained disinvestment. Land use and zoning practices, as well as steering and predatory lending practices, contributed to ongoing residential segregation within these neighborhoods. The residential segregation patterns that were established in San Antonio decades ago are still evident today (Walter et al., 2017).

Demographic Overview

San Antonio is the seventh largest city in the nation, with a population approaching 1.5 million residents. Despite being one of the least dense cities, San Antonio is one of the most populous because of its expansive layout and vast size. San Antonio has seen continuous population growth for decades, and in 2022 it ranked among the top five cities in the country for the largest numeric population gain, falling closely behind Fort Worth, Texas and Phoenix, Arizona. The median age of San Antonians is lower than that of Texas and the nation, with a smaller proportion of older individuals. Despite San Antonio being a relatively young community, it struggles with higher rates of poverty compared to both Texas and the nation. The poverty rate in the city is 18.7%, which is notably higher than in Texas (14.0%) and the United States (12.6%). In San Antonio, less than 30% of the population holds a bachelor's degree or higher, which is lower than both the state of Texas and the national average (American Community Survey 1-Years Estimates, 2022).

Originally the commercial hub of the southwest, San Antonio's Military City USA nickname comes from the host of civilian and military government jobs produced by the Joint Base San Antonio, which is the second largest employer in the city. The largest employer is H-E-B, a grocery chain, that employs over 150,000 residents. Healthcare and bio sciences, including the University of Texas Health Science Center, is another major employer. The leisure and hospitality sector are also a major source of employment in San Antonio, which sees a steady stream of tourists from all corners of the globe visiting iconic sites such as the Alamo, San Antonio Missions (a UNESCO World Heritage Site), and the vibrant River Walk, where culture, dining, shopping, and nightlife converge (Caballero & Shea-Owen, 2024).

Approximately 75% of San Antonio's population consists of people of color and this percentage is on the rise. All racial and ethnic groups in San Antonio, except the white population, have seen growth in the past decade. Two-thirds of the population identifies as Hispanic or Latinx. Despite San Antonio's racial and ethnic diversity, there are significant racial inequities in the city that are evident in many domains such as health, education, and employment. Students of color are more likely to be suspended or expelled from public school. Unemployment is highest for American Indian or Alaska Natives in the city. Households of color fall behind white homeowners in terms or ownership rates. Hispanic/Latinx are two times as likely as whites to be without health insurance (City of San Antonio, 2019).

The median household income in San Antonio is around $60,000, which is lower than Texas and the U.S., both of which have median incomes greater than $70,000. Figure 13.1 illustrates the significant disparity in income between homeowners and renters. In the past two decades, there has been a growing, and persistent, gap in the income levels of renters and homeowners. By 2022, homeowners earned nearly twice as much as renters who earned roughly $40,000 a year. Approximately 17.6% of San Antonians earn an annual income of

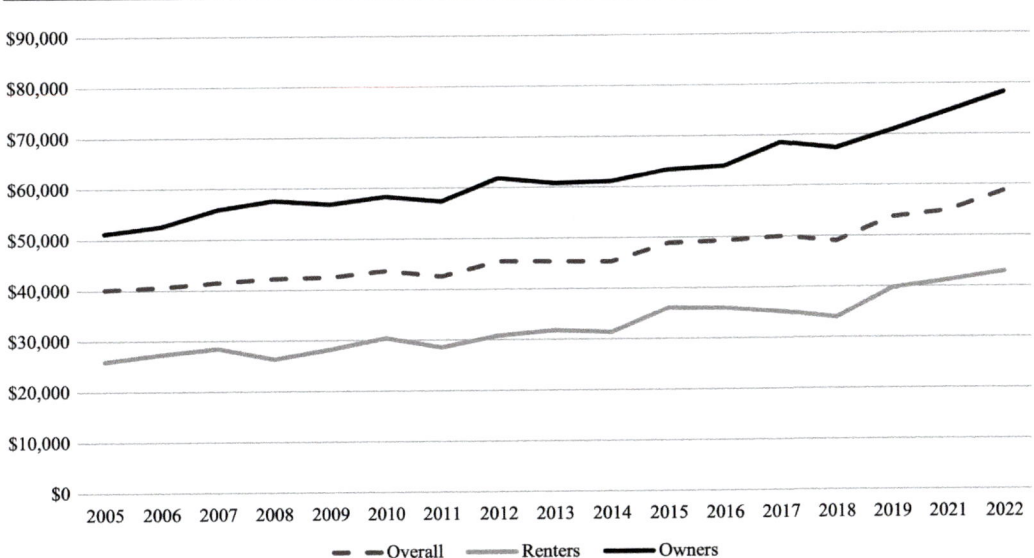

Figure 13.1 Median Annual Housing Income in San Antonio
(Source: ACS 1-Year Estimates, 2005–2022)

$13,465 or less which is below the official poverty measure. Of the largest cities in Texas, San Antonio has the highest proportion of households that earn less than $15,000 per year. Households of color comprise the majority of residents with household incomes below the poverty line (City of San Antonio, 2022). It is essential to consider low household incomes in the city when discussing broader issues of housing affordability and policies to support low-income households.

Housing Stock

The City of San Antonio has 612,031 housing units, with a 91% occupancy rate. Of these units, 52% are owner-occupied and 48% are renter-occupied. The majority are single-family homes, accounting for 59% of the total housing stock. Multi-family units make up 31% of the housing stock, while townhomes, duplexes, triplexes, and fourplexes comprise 9%. Over 30% of the housing stock has been built in the last two decades and slightly less, around 28%, is more than 50 years old (American Community Survey 1-Year Estimates, 2022).

According to Figure 13.2, the median home value in San Antonio has steadily increased over the last two decades, mirroring trends seen in many other cities across the United States. In 2005, the median home value was $89,800, but by 2022, it had increased to $230,700, representing an absolute price increase of 157%, or 5.7% per annum. The increase in home values has far exceeded the general level of inflation.

Homeowners aren't the only households that have experienced rising housing costs. As shown in Figure 13.3, the actual rents have increased from $649 per month in 2005 to $1,234 by 2022, an increase of over 90%. After adjusting for inflation, rents have increased by roughly 25%, which places significant pressure on the household budgets of renter households.

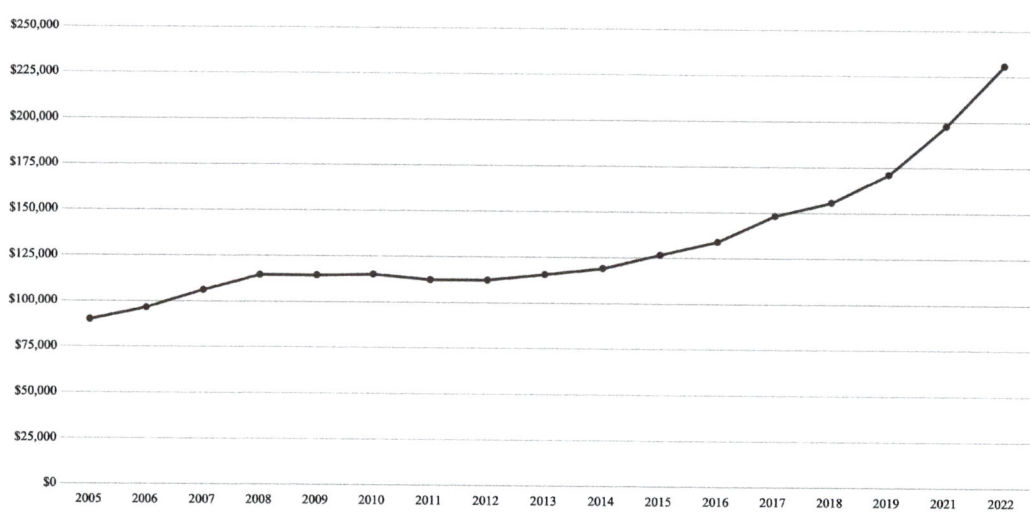

Figure 13.2 Home Value in San Antonio
(Source: ACS 1-Year Estimates, 2005–2022)

Rising housing costs have left homeowners and renters in San Antonio cost burdened. Renters at or below the area median income face significant housing affordability challenges. As highlighted in Table 13.1, a large proportion of these renters are considered cost burdened, meaning they spend more than 30% of their household income on housing expenses. Among the poorest households in San Antonio, three quarters experience severe cost burden (when housing costs exceed 50% of household income). Around 25% of moderate-income earners – those earning between 80 – 100 of AMI – are cost burdened. Nearly 60% of households between 51 and 80 AMI experience housing cost burden, while for those at, or below, 50% AMI, housing cost burden is a fact of life for the vast majority of households.

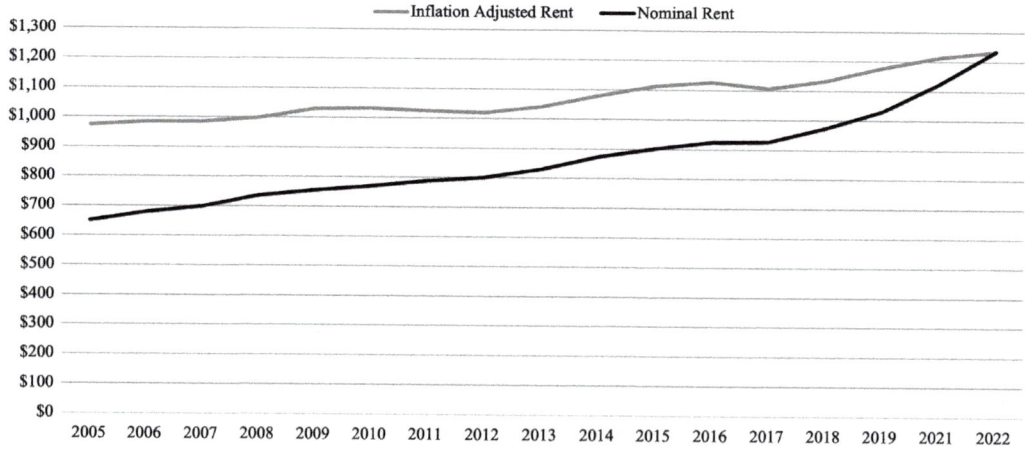

Figure 13.3 Nominal and Inflation Adjusted Median Rent
(Source: ACS 1-Year Estimates, 2005–2022)

Table 13.1 Proportion of Renter Households Cost Burdened and Severely Cost Burdened

	Cost burdened	Severely cost burdened
Extremely low income (below 30% of AMI)	88%	75%
Very low income (30–50% AMI)	87%	43%
Low income (51–80% AMI)	59%	12%
Moderate income (81–100% AMI)	26%	2%

(*Source*: National Low Income Housing Coalition GAP Report, 2022)

In addition to the increasing costs of housing, low wages are another factor contributing to the financial strain experienced by many households. For instance, in San Antonio, the fair market rent (the 40th percentile for gross rent of average quality units) for a two-bedroom apartment in 2023 was $1,282. To afford this rent without being cost burdened, a household would need to earn an annual income of $51,280. This is equivalent to working 3.4 full-time jobs at the federal minimum wage. To afford a two-bedroom apartment at the fair market rent, an hourly wage of $24.65 is required in San Antonio (National Low Income Housing Coalition, 2023).

The mismatch between household earnings and housing expenses has resulted in a shortage of affordable housing units in the City of San Antonio for households earning less than 80% of the area median income. For low-income households (earning 51–80% of AMI), there are only 92 affordable homes available for every 100 renter households. This number decreases to 45 units for very low-income households (earning 31–50% of AMI) and drops even further to 25 units for extremely low-income households (below 31% of AMI). This shortage leaves a deficit of over 80,000 units for households under 50% of AMI (National Low Income Housing Coalition, 2022). In response to the affordable housing needs in the city, the following sections outline the federal, state, and local response.

Federal Housing Supports

Affordable housing in San Antonio is supported by the federal government through various programs such as housing assistance, grants, mortgage insurance, and tax credits. The primary agencies that oversee this support are the public housing authorities in Texas. However, both the State of Texas and local jurisdictions also receive direct assistance. In San Antonio, the two public housing authorities, Opportunity Home San Antonio and Housing Authority of Bexar County, are the main agencies responsible for administering federal housing assistance programs. While Opportunity Home San Antonio focuses within the city limits, the Bexar County Housing Authority serves other municipalities within the county and unincorporated Bexar County. There is some overlap between the two agencies. As an example, both agencies administer the Housing Choice Voucher program, with residents living both within and outside of their primary service areas.

Opportunity Home San Antonio is the largest provider of affordable housing in the city. This local quasi-government agency is the main entity in the city responsible for administering federal housing assistance programs. With an annual budget of around $200 million and real estate assets exceeding $500 million, the agency provides housing assistance to more than 62,500 San Antonians through its Public Housing, Housing Choice Voucher, and mixed-income housing programs. Opportunity Home San Antonio not only offers housing, but also provides a range of supportive programs including job readiness, life skills, parenting programs, and educational opportunities (Opportunity Home San Antonio, 2024).

The two main federal grant programs that provide housing assistance in San Antonio are Community Development Block Grants and HOME Investment Partnerships Program. Both housing authorities, and Bexar County and the City of San Antonio, receive funding directly from these programs. The Federal Housing Administration Mortgage Insurance program also plays a role, offering mortgage insurance to lenders to provide flexible down payment requirements and credit guidelines for low-income residents in San Antonio. Additionally, the Federal Housing Finance Agency administers programs such as the Affordable Housing Program and Duty to Serve, which provide funding for the construction or renovation of affordable housing units.

Low Income Housing Tax Credit projects are widespread in the city of San Antonio, in which Opportunity Home San Antonio is actively involved alongside other for-profit and non-profit housing providers. Tax credits are allocated by the Texas Department of Housing and Community Affairs, the state housing finance agency, through a competitive application process for 9% credits and a non-competitive process for 4% credits. In 2024, the agency had $15,000,000 in 9% tax credits available for allocation, but received requests totaling over $50,000,000 (Texas Department of Housing and Community Affairs, 2024).

How does federal housing support affect affordability in San Antonio? With multiple funding sources and programs at play, it can be challenging to fully understand the impact. Additionally, many developments have multiple funding sources further complicating the ability to track impact. According to HUD's Picture of Subsidized Households database, in 2022 the number of subsidized units available supported by HUD was 23,735. This means approximately 3.9% of San Antonio's housing stock is supported by federal housing support from HUD which includes Public Housing, Housing Choice Vouchers, other project-based programs like Project Based Section 8, and multi-family housing programs such as 811 and 202 (all discussed in Chapter 8 on supply side housing). Given that nearly 19% of all San Antonians live below the federal poverty line, the inadequacy of federal housing supports is evident. Roughly speaking, there are about 15% of households in San Antonio that experience poverty yet receive no federal housing support. This places significant pressure on state, county, and local government to support these households.

However, this is likely an undercount because it doesn't capture all programs mentioned above such as tax credit units, HOME funds, and some other non-HUD federal loan guaranteed programs for example. The National Preservation Database incorporates HUDs programs and programs that are administered by other federally agencies like IRS tax credits. For all active property subsidies through 2022, the National Preservation Database indicates 30,158 federally supported units. This figure provides a more thorough overview of federal assistance, revealing that approximately 4.9% of the housing inventory in San Antonio receives support from federal programs. This percentage is significantly lower than what is observed in Chicago and slightly lower than the national average. While beneficial, additional housing support is required to tackle the shortage of affordable units in San Antonio.

State and Local Housing Supports

State and local housing assistance is even more challenging to track than federal support due to the numerous and diverse sources of funding. There are two primary programs administered at the state level that support households. First, there is the federal tax credit program that has been described earlier. These credits are allocated to communities and developers by the Texas

Department of Housing and Community Affairs. Second, the state housing trust fund allocates money for housing to communities throughout the state of Texas. The trust fund is funded with general revenue from the state and generates approximately $5 million annually (Texas Department of Housing and Community Affairs, 2024). Because the fund supports state level activities – such as providing loans and grants for housing initiatives that assist low-income households throughout the state – it is difficult to determine what share of those funds support low-income households in San Antonio.

To supplement funds from both national and state housing trust funds, San Antonio established its own. The San Antonio Housing Trust, a local non-profit organization, manages a housing trust fund that was established in 1991 through a partnership with the City of San Antonio. While the fund does not have a designated funding source, it was initially seeded with contributions from the city and other donors. These funds are used to develop new housing units and to preserve existing units for residents in San Antonio earning below 60% of the AMI. Since its inception, the trust has facilitated the creation or preservation of over 11,000 housing units in the city and distributed more than $10 million in grants and loans (San Antonio Housing Trust, 2024).

The City of San Antonio's Neighborhood & Housing Services department offers a variety of housing support programs such as eviction prevention, rental assistance, homeownership programs, home repair, and residential relocation assistance. Additionally, the department provides programs that offer legal advice, property tax relief, tenant/landlord remediation, an accessory dwelling unit program, and affordable housing bonds. The department also organizes housing related events and manages a housing commission that advises the city on housing goals, initiatives, and policies. These efforts are all part of the Strategic Housing Implementation Plan developed in 2020. Mayor Ron Nirenberg initiated the Mayor's Housing Policy Task Force in 2017. The recommendations of the task force formed the basis of the city's 10-year housing plan.

Annually, the Housing Commission and Department of Neighborhood & Housing Services, both of the City of San Antonio, collaborate to create a report detailing the advances made in the city's 10-year housing plan. A major milestone was the approval of the city's first housing bond in 2022, totaling $150 million. Within a year, the city successfully distributed funding for 4,833 homes, with 81% of the units designated as affordable. Progress on the construction and preservation of new homes stands at 37% of the goal, with 10,400 units completed out of the 10-year target of 28,094. In addition to housing developments, the city implemented various programs such as an accessory dwelling unit design competition, the Green and Health Homes Initiative, and Housing Base, an online platform assisting residents in finding affordable housing options (City of San Antonio, Neighborhood & Housing Services, 2023).

Numerous non-profit organizations, private sector entities, foundations, advocacy agencies, and grassroots efforts are actively involved in promoting affordable housing in San Antonio. While not nearly an exhaustive list, Table 13.2 highlights the key players that are dedicated to expanding housing opportunities for low-income households in the city.

In addition to partnering with and supporting local organizations to promote affordable housing, the city is also pursuing changes to its regulatory approach to housing. Examples include the one-stop housing center, an online portal created by the City of San Antonio to coordinate a community-wide housing system. The online portal, in partnership with a local non-profit, was designed to provide a centralized location for all things related to housing like rental assistance, temporary housing assistance, homebuyer education courses, and legal assistance. The city is

Table 13.2 Organizations that Promote Affordable Housing in San Antonio

Organization name	Role	Impact
LISC San Antonio (non-profit community development organization)	Supports housing development and preservation	Invested over $16 million since 2016
San Antonio Habitat for Humanity (Christian-based non-profit)	Builds affordable single-family homes with the families that are going to buy them	Built over 1,300 Habitat homes
Housing First Community Coalition/Towne Twin Village (non-profit organization)	Builds new permanent supportive housing communities for older adults	200 new permanent supportive units
NRP San Antonio (for-profit multifamily developer)	Builds new apartment homes through public-private partnerships and with tax credits	61 affordable, mixed income rental communities (13,369 units)
Neighborhood Housing Services San Antonio (non-profit CHDO and CDFI)	Develops and rehabs affordable homes, provides financial assistance and counseling services	Built 325 new homes, rehabbed 200 homes, 14,000 loans, counseling 300+ clients per year

(Source: LISC San Antonio, San Antonio Habitat for Humanity, Towne Twin Village, NRP San Antonio, Neighborhood Housing Services San Antonio, 2024)

also updating the Unified Development Code to remove barriers and regulatory obstacles to housing development. San Antonio has also adopted a public communication strategy to underscore and raise awareness of the importance of affordable housing in the city.

Looking Ahead

The City of San Antonio stands out as an example of a proactive approach to addressing the affordable housing crisis. Through a well-coordinated plan, the city has brought together various partners to work toward common goals. This collaborative effort has led to the development of new affordable housing units, as well as the rehabilitation and preservation of existing ones. Additionally, homebuying counseling programs have expanded opportunity for households who desire to purchase a home and assistance to help renter households maintain stable housing. The city has a clear and transparent process to monitor these programs which enhances its effectiveness and accountability. The success of these efforts helps build community support for affordable housing which was evident in the passage of the $150 million affordable housing bond.

San Antonio's initiative, Reframing Housing, is noteworthy. The campaign aims to educate residents on the importance of home as the center of life and a fundamental need, as well as the cornerstone of economic mobility that benefits society as a whole. The campaign dispels the misconception that affordable housing is only for undeserving individuals that receive government assistance, and instead highlights that these programs support older adults on fixed incomes, provide rent and mortgage assistance for essential workers like teachers, and assist households that face significant hardships such as the loss of a primary breadwinner. The campaign emphasizes that rising housing costs leave little room for the typical household to afford other essential expenses like food and medical care. This outreach helps to build broad support for greater investments in affordable housing, which is necessary to bring about meaningful change.

Although the City has made progress in addressing housing affordability, criticisms and problems persist. Many low- and moderate-income households remain housing cost-burdened. Most of the affordable housing initiatives often lack the necessary deep subsidies required to house extremely low-income households earning under 30% of AMI. The right to housing for all San Antonians as a stated city goal remains an unmet need. Homelessness continues to be a pressing concern, with a 7% increase over the last year. The 2024 point-in-time count estimates that 3,372 individuals experienced homelessness in the San Antonio region on a given night, with 888 people living unsheltered (Close to Home, 2024).

To respond to the ongoing challenges, especially for those households with the lowest incomes, San Antonio implemented a basic income pilot in 2020, which provided 1,000 households with upfront cash assistance of $1,908 and subsequent support of $400 a quarter for eight quarters until January 2023. Evaluations of the program indicate that the funds were used primarily for housing, utilities, food, clothing, transportation, and debt repayment (consistent with the findings of basic income pilots in other jurisdictions). Some recipients also used the money to support career or educational goals. While the pilot did not create long-term assets, it did provide temporary assistance for essential needs (Joyner et al., 2024). Unless there is a significant change in housing policy, like the Housing Choice Voucher program becoming an entitlement, a more robust basic guaranteed income is another option to effectively tackle the affordable housing crisis in cities across America.

References

Caballero, N., & Shea-Owen, E. (2024). *Top industries and employers in the San Antonio metro area. SATX Today*. Retrieved from https://satxtoday.6amcity.com/city-guide/work/top-industries-employers-san-antonio-tx

Caine, I., Walter, R., & Foote, N. (2017). San Antonio 360: The rise and decline of the concentric city 1890–2010. *Sustainability, 9*(4), 649.

City of San Antonio. (2019). *Racial equity indicator report*. Retrieved from www.sanantonio.gov/Portals/0/Files/Equity/IndicatorReport.pdf

City of San Antonio. (2022). *Status of poverty in San Antonio*. Retrieved from www.sanantonio.gov/Portals/0/Files/HumanServices/Poverty/2022PovertyReport.pdf

City of San Antonio, Neighborhood & Housing Services. (2023). *Housing Commission 2023 annual report*. Retrieved from www.sanantonio.gov/Portals/0/Files/NHSD/Coordinated%20Housing%20Web-page/SHIP/2023%20Annual%20Report_ENGLISH.pdf?ver=2024-01-23-095409-693

Close to Home. (2024). *State of homelessness report*. https://sanantonioreport.org/wp-content/uploads/2024/05/2024-State-of-Homelessness-Report-Close-to-Home.pdf

Dimmick, I. (2022*). Cash without conditions: San Antonio's experiment with guaranteed income*. San Antonio, TX: San Antonio Report.

Fairbanks, R. B. (2022). The Texas exception: San Antonio and urban renewal, 1949–1965. *Journal of Planning History, 1*(2) 181–196.

Joyner, K., Denton, M., & Charkraborty, D. (2024). *Social determinants of health study*. Retrieved from www.uptogether.org/wp-content/uploads/2024/01/UTSA_UpTogether_Methodist-Healthcare-Ministries-Social-Determinants-of-Health-Study-2-1.pdf

National Low Income Housing Coalition. (2023). *The Gap Report*. State of Texas. Retrieved from https://nlihc.org/gap/state/tx

National Low Income Housing Coalition. (2023). *Out of reach*. The High Cost of Housing. Retrieved from https://nlihc.org/oor

Opportunity Home San Antonio. (2024). *About Opportunity Home*. Retrieved from https://homesa.org/housing/about-homesa/

San Antonio Housing Trust. (2024). *Who we are, what we've done*. Retrieved from https://sahousingtrust.org/about-us/ and https://sahousingtrust.org/about-us/

Texas Department of Housing and Community Affairs. (2024). *Competitive (9%) housing tax credits.* Retrieved from www.tdhca.texas.gov/competitive-9-housing-tax-credits

Texas Department of Housing and Community Affairs. (2024). *HTF funding sources and background.* Retrieved from www.tdhca.texas.gov/htf-funding-sources-and-background

Walter, R. J., Foote, N., Cordoba, H. A., & Sparks, C. (2017). Historic roots modern residential segregation in a southwestern metropolis: San Antonio, Texas in 1910 and 2010. *Urban Science, 1*(2), 19.

Zelman, D. L. (1983). Alazan-Apache courts: A new deal response to Mexican American housing conditions in San Antonio. *Texas State Historical Quarterly, 87,* 123–150.

Chapter 14

Seattle, WA

Overview

Seattle, Washington – also known as the Emerald City – serves as the economic hub of the Puget Sound region. Wedged between the Olympic and Cascade mountain ranges, and bordering the blue waters of the Puget Sound, Seattle's natural beauty is virtually unmatched among large cities in the United States. One of the many reasons that Seattle continues to be a desirable location for the millions of people who have moved to this region over the last two decades is its beauty and the easy access it provides to a range of outdoor recreational activities.

The Puget Sound region is the ancestral home of the Coast Salish people, and Seattle got its name from the chief of the Suquamish and Duwamish people. White settlers from the east arrived, in what would become Seattle, in 1851. The economic cornerstone of this young city was the timber industry given the abundant trees in this region. A large mill was established in what is currently downtown Seattle with easy access to Elliott Bay. The mill produced much of the lumber that was used to construct homes and buildings in San Francisco in the middle to latter half of the nineteenth century.

The arrival of the Northern Pacific Railway in 1870 and the Great Northern Railway in 1893 linked Seattle to the rest of the country. The discovery of gold in Alaska and the Yukon Territory turned Seattle into a boom town as prospectors used Seattle as a staging area to begin their journey in search of gold. The boom associated with the gold rush initiated a cycle of booms and busts throughout the life of Seattle. Much of the economic and housing history of Seattle can be directly linked to a series of booms, and subsequent busts.

The next source of economic prosperity and growth in the region came from the aviation business, in particular, the Boeing Company. World War II stimulated demand for Boeing's airplanes, but the end of the war pushed the company, and the region that it called home, into an economic decline. Boeing re-emerged in the late 1950s with the creation of its 707 airplane which prompted another period of growth and prosperity for the company. In the midst of this boom, Seattle hosted the World's Fair in 1962 in which it presented itself as a futuristic city. Seattle's iconic Space Needle was created for this event and continues to mark the city's skyline to this day.

A decade later, Seattle, again, found itself in a period of economic decline. The period is known as the Boeing Bust; in 1967 Boeing employed over 100,000 people and within four years that number had fallen to just under 39,000 (Lange, 1999). At the peak of the Boeing Bust, the unemployment rate in Seattle was nearly triple that of the national average. In response, two local real estate agents, seeking to have some fun with the local pessimism, rented a local billboard near Sea-Tac International Airport that read, "Will the Last Person Leaving SEATTLE – Turn Out the Lights."

DOI: 10.1201/9781003356585-18

The next boom cycle in Seattle was not driven by lumber or aviation, but rather technology. In 1980, Bill Gates and Paul Allen formed Microsoft in nearby Redmond, Washington. Microsoft changed the world of personal computing and had a profound impact on the Puget Sound region. Just over a decade later, Jeff Bezos launched Amazon in Seattle. The success of Amazon brought a surge in population, tremendous wealth, and a radically altered built environment to the City of Seattle. Among the many booms that Seattle has experienced, this tech driven boom has arguably had the most significant impact on the region.

One of the byproducts of the technology fueled boom of the last couple of decades in Seattle has been a dramatic increase in population and wealth. A significant outcome associated with economic booms is a rapid increase in housing prices. Seattle has followed this typical pattern. While the increase in housing prices has provided significant wealth benefits for homeowners, it has created substantial challenges for lower income households. The boom has produced high housing cost burdens, displacement especially among households of color, and a crisis of homelessness.

Demographic Overview

In 1980, when Microsoft was formed, the population of Seattle was just under 500,000; by 2020 it had grown to nearly 740,000. In the decade of the 2010s alone, the population of Seattle grew by over 21%, which was largely attributable to the Amazon boom. The broader region has also experienced dramatic growth. The Seattle Metropolitan Statistical Area (MSA) had about 1.8 million residents in 1980 and by 2020 it had nearly doubled to over 3.4 million people. Few regions in the country grew this quickly.

With this growth came a dramatic increase in income and wealth. In 2022, the median household income in Seattle was over $115,000 and the average was $167,000, which speaks to the significant percentage of high-income households in the city. For the nation as a whole, the equivalent figures were $75,000 (median) and $105,000 (average). Seattle has incomes that are over 50% greater than the national average. According to data from the St. Louis Federal Reserve Bank, the Seattle MSA has the fourth highest gross domestic product (GDP) per capita. GDP is a measure of economic output in a given geography and this measure highlights the significant wealth and prosperity of this region.

Consistent with intuition, Seattle has a lower poverty rate than many large cities. The combination of robust employment with one of the nation's highest minimum wages, has resulted in a city-wide poverty rate of only 10.1%, compared to the national average of 12.6%. Seattle's poverty rate is even lower than other wealthy cities, including San Francisco and Boston. But even though Seattle enjoys relatively low poverty, this figure still highlights a conspicuous challenge: nearly 1 in 10 people live below the federal poverty line in one of the most expensive cities in the nation. This combination presents a significant challenge for low-income households.

Breaking down the population of Seattle by housing tenure yields other notable observations. As highlighted in Figure 14.1, consistent with national averages, homeowners have much higher household incomes than do renter households. Since 2017, there has been a pronounced divergence as homeowner incomes have accelerated more rapidly, while renter incomes have stagnated. This graphic provides additional evidence that Seattle has become a city of "haves" and "have nots."

Finally, Seattle is home to a large, and persistent, population of people experiencing homelessness. On a single night in 2024, the Seattle-King County region estimated that 16,385 people experienced homelessness – either living in a homelessness shelter or residing in vehicles or on

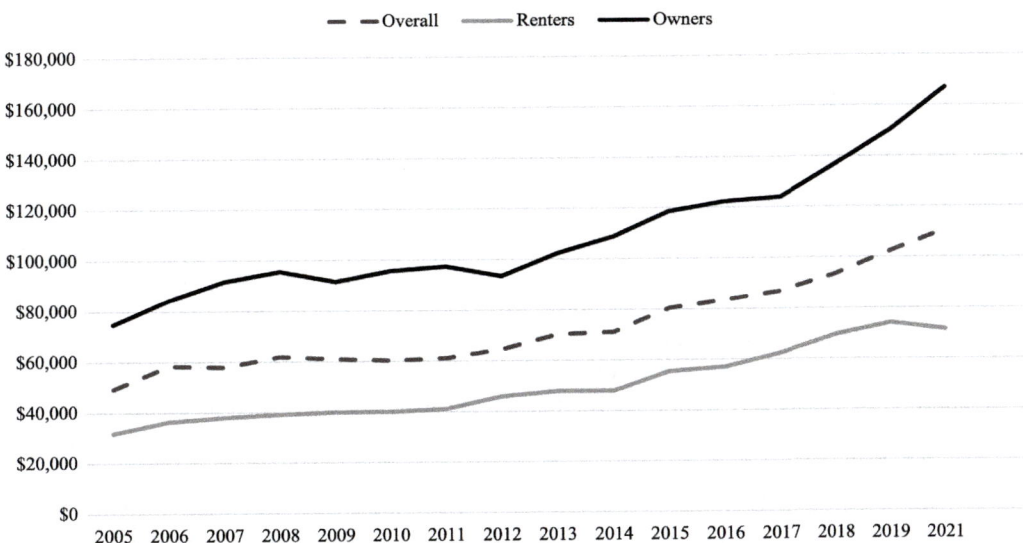

Figure 14.1 Median Household Income in Seattle, by Housing Tenure
(Source: American Community Survey 1-Year Estimates)

the street. This region is home to one of the largest per capita rates of homelessness in the nation. While tragic, it should come as no surprise that Seattle is home to such a large population of people experiencing homelessness given the substantial challenges of low-income households to find housing in a challenging housing market like Seattle. Abundant research has drawn a causal link between the unaffordability of housing and high rates of homelessness (Colburn & Aldern, 2022). Public opinion polling highlights that homelessness has become one of the major areas of concern for voters in Seattle and throughout the State of Washington.

Housing Stock and Market Overview

According to data from the U.S. Census American Community Survey, in 2022 Seattle was home to 367,119 households, 56% of which were renter households. These households have just shy of 400,000 housing units from which to choose. These figures imply that there are 30,609 vacant housing units in Seattle, or 7.7% of all housing units. By comparison, the comparable figure in Chicago is 9.9%. For a city that struggles to house many of its people, this 7.7% vacancy figure seems surprising. But there are a couple of points that are worthy of attention. First, the total vacancy number that includes a wide variety of vacant units that doesn't necessarily have an impact on households that are seeking housing (either ownership or rental). Also, vacancy rates have increased in the City of Seattle since the pandemic. The rental market vacancy rate (which is different than the overall vacancy rate) rose from an extremely low 3.8% in 2018, to a more moderate 5.9% in 2022, and the overall vacancy rate has increased from 6.3% to 7.7% over the same period.

Table 14.1 breaks down the 30,609 vacant housing units in the City of Seattle in 2022. As Table 14.1 highlights, only about a third of all housing vacancies are classified as "for rent" that prospective renters could potentially lease. Therefore, options remain limited. But, slower population growth since the pandemic and an increase in housing production has created

Table 14.1 Vacant units in the City of Seattle in 2022

For rent	13,225
Rented, not occupied	3,392
For sale only	1,664
Sold, not occupied	770
For season, recreational, or occasional use	4,613
Other vacant	6,695
For migrants	250
Total vacant units	30,609

more housing options for renter households in Seattle compared to the exceedingly tight market conditions that existed in the years prior to the Covid-19 pandemic.

In addition to a generally tight housing market, the second reason why accessing housing in Seattle remains difficult is its cost. The region remains one of the highest cost locations in the country. While the extra vacancies do provide more options for people, high rental prices remain a challenge. According to the American Community Survey, the median gross rent in Seattle in 2022 was nearly $1,900 per month, which was down from over $2,000 in 2019. Even the slightly lower 2022 figure is nearly 50% higher than the national average. Rents in Seattle are higher than Los Angeles and slightly lower than those found in San Diego and San Francisco. Therefore, even with slightly higher vacancies in the early 2020s, high rent levels present significant challenges for low-income households seeking housing in Seattle.

While renting in Seattle is very expensive, so is homeownership. The median value of owner-occupied housing in the City of Seattle in 2022 was $924,200, higher than both Los Angeles and San Diego. Among west coast peers, only San Francisco was higher at $1,343,700. It is also important to underscore what an outlier coastal cities are in terms of housing affordability. As a reminder, the same figure in Chicago was only $313,300.

The high housing costs, combined with relatively high incomes, still produces highly unaffordable housing in Seattle. According to the ACS in 2022, almost 46% of all renters in Seattle spent more than 30% of their household income for housing expenses. But this figure is actually less than the national average which is 52%. The higher incomes in Seattle help to explain the modestly lower levels of housing cost burden among renters. But what is missing in the coarse analysis of housing cost burden is how the high absolute levels of rent present significant challenges for low-income households. The higher average wages earned by households in Seattle are little comfort to low-income households that face high rents with limited income.

An obvious question, therefore, is what explains the challenging housing market conditions that we observe in Seattle and other coastal cities? Is it population growth? Surely this is part of the story. Is it the high wages and wealth earned by employees in the technology industry? Again, yes, this has a clear effect on housing costs. But these factors are only part of the story; a lack of housing supply has exacerbated the housing crisis that we observe in Seattle and other similar cities.

A key concept in the field of economics is supply elasticity. The price elasticity of supply measures the percentage change in the supply of a good for a given percentage change in the price of that good. Urban economists have applied this notion to the field of housing. The question for a city or region is what happens to housing supply when the price of housing changes? In other words, when people move to a community, and prices rise, is more housing constructed? The two primary inputs into determining how elastic a city's supply response will be are the regulatory environment and its topography. Locations with challenging topography and tight housing regulations, will experience inelastic housing supply ... meaning they don't build much housing when prices rise. The combination of rapid population growth with inelastic housing supply is a recipe for a housing affordability crisis. And that is what has transpired in many of our coastal cities.

Because Seattle is land locked, it lacks the opportunity for sprawl that is observed in San Antonio and many other fast-growing locations in the Sun Belt. Therefore, there is tremendous pressure to maximize the land that does exist for residential uses. The result tends to be high land prices which increases the cost of housing. In this situation, the high land prices place significant pressure on zoning and land use. Given the scarce land, cities such as Seattle must maximize the housing capacity on each parcel rather than relying disproportionately on single-family zoning.

In Gregg's book, *Homelessness is a Housing Problem: How Structural Factors Explain U.S. Patterns*, he and his co-author, Clayton Page Aldern, highlight the relationship between the demand for housing (population growth) and its supply (as measured by supply elasticity). Figure 14.2 comes from that book and shows different clusters of cities in the U.S. based on these two variables. All three cities that serve as case studies in this book are represented in the figure.

Figure 14.2 highlights a few important points. First, the combination of rapid population growth combined with a low housing supply elasticity, as is found in Seattle, is problematic. The very low rental housing vacancy rates (3.1%) found in Seattle are evidence of the mismatch between the demand for housing and its supply. Second, population growth alone does not necessitate a housing crisis. In the upper right quadrant, there are a number of Sun Belt cities,

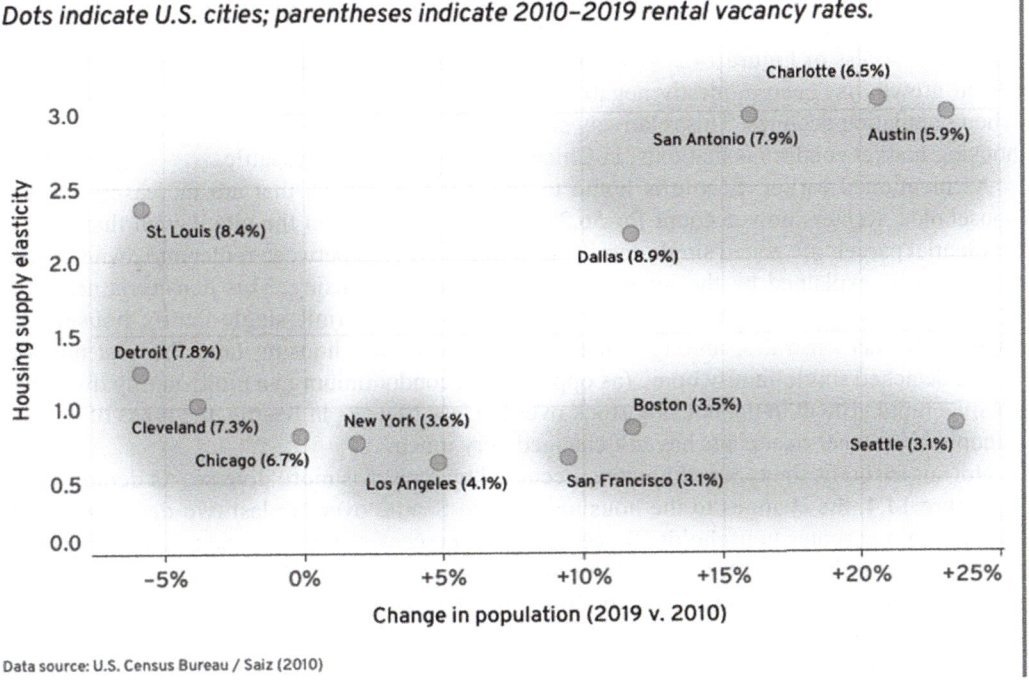

Figure 14.2 Population Growth Versus Housing Supply Elasticity
(Source: Colburn & Aldern, 2022)

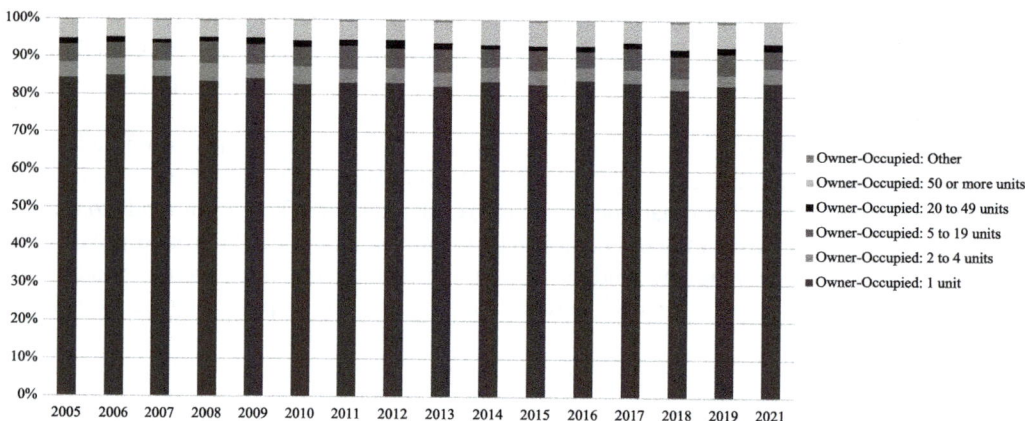

Figure 14.3 Type of Owner-Occupied Housing in Seattle, 2005–2021
(Source: American Community Survey 1-Year Estimates)

including San Antonio, that experienced robust population growth during the decade of the 2010s, but didn't have the same experience of rapid housing cost appreciation. Why? The reason is their relatively elastic housing supply that allowed them to build sufficient housing to accommodate the significant growth in those cities. Finally, inelastic housing supply alone, does not create a housing crisis. Our first case study city, Chicago, highlights this phenomenon. Because Chicago borders Lake Michigan, it has topographical constraints like coastal cities, but it hasn't experienced the same housing market pressures. The reason for this split can be found on the horizontal access of Figure 14.2 – Chicago has not experienced population growth. In fact, its recent growth has been modestly negative. Therefore, low supply elasticity is of less concern when population declines. This relatively simple graphic helps to explain the radically different housing market conditions that exist in Chicago, San Antonio, and Seattle.

As mentioned earlier, Seattle is home to 367,000 households, that are increasingly renter households. Renters now account for 56.2% of the households in the city. Given that 70% of residential parcels are zoned single-family as of 2022, the split between renter and owner households can be explained by the increased density of rental housing. This density is necessary given that such a significant percentage of parcels only permit single-family housing. As Figure 14.3 demonstrates, among households that own their housing unit, the vast majority own a detached single-family home (as opposed to a condominium in a multi-family dwelling). Despite rapid growth in the housing stock over the last 20 years in Seattle, the types of housing occupied by owner-occupants haven't changed very much.

Not surprisingly, the types of housing occupied by renters is more diverse. As demonstrated in Figure 14.4, the changes to the housing stock in Seattle over the last two decades is more evident among renter households. Since 2005, the percent of renters living in buildings with four or fewer units has fallen from about 40% to roughly 30%. This speaks to the increase in large apartment buildings in Seattle. In 2005, just over 20% of renters lived in buildings with 50 or more units, and by 2021 that number had grown to about 35%. One of the problems with the rental housing stock is how few of these units are family-sized. The vast majority of rental units are studios and one bedroom which limit housing access for larger households with children. Of the over 206,000 renter households in the City of Seattle, nearly 89% of these households have only one or two people living in that rental unit. Family-sized renter housing units in Seattle are scarce.

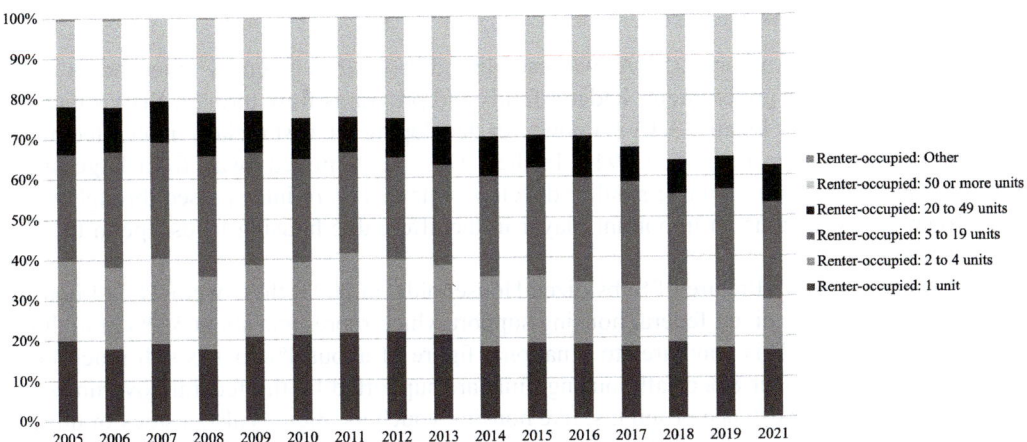

Figure 14.4 Type of Renter-Occupied Housing in Seattle, 2005–2021
(Source: American Community Survey 1-Year Estimates)

It is wrong to suggest that Seattle hasn't created new housing as its population grew. In the ten years between 2012 and 2022, the housing stock in the City of Seattle grew by 18.6% and the population grew by just over 18%. The obvious question, therefore, is why Seattle has such a housing shortage. There are two ways to answer that questions. First, there was likely a housing shortage that existed prior to 2012 that has persisted as housing production has kept pace with population growth over the last ten years. Second, a broader geographic focus highlights the acute housing shortage throughout the region. According to research from the Puget Sound Regional Council, as of 2022, the broader region has a current deficit of 46,000 housing units (Puget Sound Regional Council, 2022).

The seeds of this housing shortage have been sown over 40 years. In the 1980s, the Puget Sound region produced more than 230,000 housing units. That production number has declined since the 1980s to just over 160,000 housing units in the decade of the 2010s – a decade of rapid population growth in the region (Puget Sound Regional Council, 2022). Given the expected ongoing increase in population in the region over the next 30 years, the Council suggests that the region will need an additional 800,000 housing units over that period. The encouraging news is that regional housing production has increased dramatically over the last five years, but to meet the regional needs for the next 30 years, 30,000 units will need to be produced each year, which is a higher level of production than even during the recent construction boom in the region. Failure to accommodate this ongoing population growth could have significant and severe consequences for affordability in Seattle and its neighboring communities.

While beyond the scope of this case study, creating the needed housing capacity in the region will require two separate, but related, efforts. First, Seattle and its suburbs must confront the reality of how housing regulations have limited housing supply. As mentioned above, exclusionary land use – single-family zoning – limits the amount of housing that can be constructed in a community. It will be difficult to create the housing that is needed without changing land use which will allow for greater housing density than is currently permissible. Second, as the region makes an historic investment in light rail throughout the region, that buildout creates the opportunity to create significant housing capacity around light rail station areas. Failure to create dense housing at those station areas would be a missed opportunity as the region seeks to meet the projected 800,000 housing unit need according to the Puget Sound Regional Council.

Federal Housing Supports

The Seattle Housing Authority (SHA) was established in 1939 as an entity independent of the federal, state, or local government. Federal funds and programs flow through SHA which administers the programs locally. SHA is known for developing the first racially-integrated public housing project, Yesler Terrace in 1941. In the 1970s, SHA moved away from large public housing developments in favor of smaller developments that were interspersed throughout the city. Today, SHA remains an important player in the affordable housing landscape in the City of Seattle.

According to HUD's Picture of Subsidized Households, in 2023, there were 21,134 housing units in Seattle that received federal housing support which represents about 5.3% of all housing units in the city. This compares to a national figure of about 4%, but is still much lower than Chicago where over 8% of all housing units are supported by the federal government. In addition to housing assistance programs, like housing vouchers, there is also a stock of housing units built with the support of the Low Income Housing Tax Credit (LIHTC) program. According to the HUD LIHTC database, in Seattle, there are roughly 29,500 affordable units built with the support of tax credits across 311 projects. These units represent roughly 7.4% of the city's housing stock. Using similar math that we did in the Chicago chapter, if we assume 30% subsidy overlap, a rough estimate of the total housing stock that is subsidized by the federal government is 10.4%, slightly lower than the 12.4% in Chicago. If the rate of overlap is higher than 30% (which it well could be in an expensive market like Seattle) the percent of total units subsidized by the federal government – via rental assistance or tax credits – would be less than the 10.4% mentioned above.

State and Local Housing Supports

Arguably, the most significant state level policy that has influenced the production of affordable housing in both the State of Washington and the City of Seattle is the state Housing Trust Fund. The fund was established in 1986 and, since that time, has invested over $2 billion to support the creation of affordable housing in the state. The program is managed by the State Department of Commerce. The state distributes funds from the trust fund to local governments, housing authorities, nonprofits, and tribes. Given the scale of the housing crisis in Washington, the legislature has increased funding for the Housing Trust Fund. In 2023, the legislature approved $400 million for the fund.

In the last couple of years, there has been a renewed interest in housing policy at the state level. The magnitude of the crisis has forced these issues on policymakers. The 2023 legislative session was known as the "year of housing" and ten different housing related bills were passed and signed into law by Governor Jay Inslee. Included in these bills was a middle housing bill that legalized duplexes and fourplexes in many neighborhoods, a bill that supported the development of accessory dwelling units, and legislation that streamlined permitting for new construction. On top of these bills, the legislature approved $1 billion for housing (which included the $400 million commitment to the Housing Trust Fund). Finally, the legislature passed a covenant homeownership bill that seeks to address systemic discrimination in the homeownership market by providing down payment assistance to Black households that have been subject to discrimination in the housing and mortgage market.

Unlike the success of 2023, 2024's legislative session was a disappointment for housing advocates. Of the over 100 housing-related bills that were considered, only a few were signed into law. Among the proposals that failed to pass were an effort to create more housing density near transit, a bill to create a State Department of Housing, and a couple of rent stability proposals. There is broad agreement that the state can, and must, do more to address housing affordability in Seattle and throughout the state.

One state level policy that has had a significant impact on the stock of housing, and particularly affordable housing, is the Multifamily Tax Exemption (MFTE) program which was established in 1995 by the Washington State Legislature. The MFTE program is statewide, but each jurisdiction has autonomy to create specific terms and conditions for how the program will operate in that jurisdiction. The local variation in MFTE programs has led to unequal results of the program. In some locations, communities have established strict affordability requirements for any developments that use the MFTE, while other jurisdictions have less strict conditions. Therefore, in some locations, MFTE is explicitly an affordable housing program, while in other locations it is a tool to encourage the development of multi-family housing more generally.

MFTE is a prime example of how multiple levels of government can be involved in similar policies. The City of Seattle has established its own set of rules around the use of the MFTE program in the city. Since inception, developers have created about 31,000 units of multi-family housing with support of MFTE and 6,300 of those units are affordable (City of Seattle, 2024a). The program has been reauthorized six times since inception and the City of Seattle is currently considering changes to the program rules as it considers its seventh reauthorization. Under Program Six (the sixth reauthorization) the program requires developers to set aside between 20 and 25% of units to be affordable in exchange for a waiver of property taxes. A key question that policymakers are wrestling with is whether the benefits of the program (the additional units of affordable housing) justify the costs of the program (the lost tax revenue).

In addition to the policies provided by the State of Washington, the City of Seattle has an active housing policymaking agenda. Over the last couple of decades, there are a couple of policies that have become fundamental components of the city's efforts to provide affordable housing. First is the Seattle Housing Levy. This levy, designed to provide funds specially set aside to support the delivery of affordable housing, was first passed in 1986. Every seven years the levy has been reauthorized based on an affirmative vote of Seattle voters. The most recent vote was held in 2023 when over 69% voted in favor of the $970 million housing levy that will be spent over seven years. The levy was funded with additional property taxes that cost the median Seattle homeowner $383 in additional property taxes per annum. The 2023 iteration of the levy represents almost a tripling of the prior levy that was approved in 2017, but given the dramatic increase in the cost of housing, it represents a modest increase in inflation-adjusted terms. The funds provided from the 2023 reauthorization will be used to fund the development of affordable housing, preserve existing affordable housing, and provide rental assistance. According to City of Seattle data, as of 2022, more than 16,000 people live in housing that has been supported by funds raised from the levy. The consistent support for the Seattle Housing Levy is tangible evidence of the recognition among Seattle voters that housing is an area of great need for the city. Figure 14.5 shows how the scale of the housing levy has changed over time as has its intended uses.

In 2015, the Seattle City Council passed the Affordable Housing linkage fee, which was the precursor to the Mandatory Affordable Housing (MHA) program. The concept of this program is to require developers to create affordable housing units in new developments (3 to 5% of units affordable for residents at 60 to 80% of area median income). In lieu of onsite units, developers can choose to pay a fee to the city which can be used to fund affordable housing elsewhere in the city. The size of the fee is based on the location of the development. The MHA program is

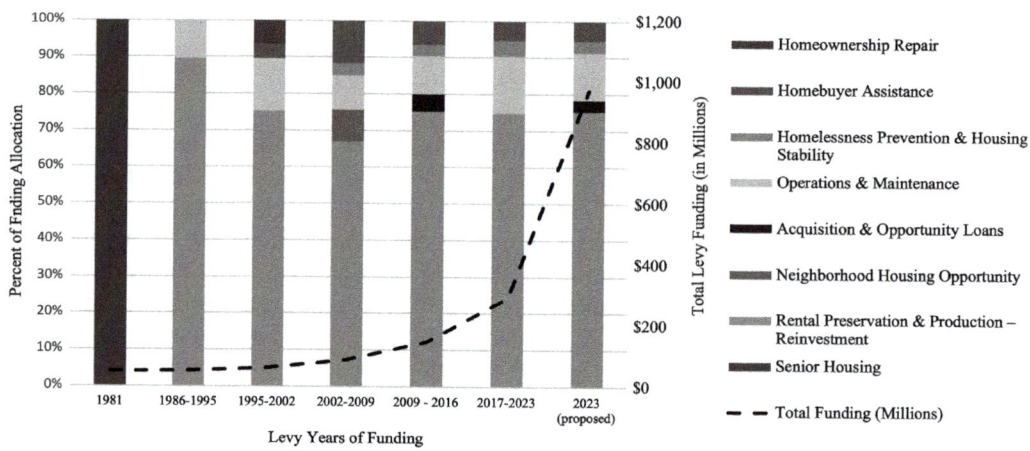

Figure 14.5 History of Seattle Housing Levy

an example of an inclusionary zoning policy that is used in many cities throughout the United States. Since inception, a smaller subset of developments has chosen the performance option in which the developer creates the affordable housing units in that development. Through 2023, there have been 56 performance projects totaling 5,086 units of which 404 units were affordable. Of the projects that chose the in lieu payment option, over $304 million has been collected pursuant to MHA and those funds have been awarded to fund affordable housing projects throughout the City of Seattle (City of Seattle, 2024b).

MHA was one element of a broader housing affordability initiative established by the City of Seattle. In 2015, an advisory committee issued 65 different recommendations to the city that could be used to address the housing affordability crisis. The effort was known as the Housing Affordability and Livability Agenda (HALA). Many of the recommendations included in HALA were related to land use and regulatory reforms that could create greater capacity for affordable housing in the city. A key element of HALA was MHA (introduced above) which was described as the "Grand Bargain." In exchange for upzones, that increased the development capacity of a given parcel of land, developers agreed to provide affordable housing units. MHA continues to be in effect in the City of Seattle, but the remaining efforts recommended by the HALA advisory committee ended four years later with a mixed report card. Researcher Alan Durning, who served on the initiative's advisory committee, wrote a thorough summary of HALA in 2020 (Durning, 2020). According to Durning, the evaluation of HALA has three components. First, HALA produced a number of notable wins, which is difficult given the knotty local politics around housing. Second, on the other hand, Durning suggests that the successes were not as notable as one might have hoped, "They're [the successes] compromised, flawed, or otherwise not really what HALA recommended. That's why the housing shortage HALA set out to cure has instead worsened" (Durning, 2020). Finally, HALA, despite its shortcomings, brought great attention to the issue of housing affordability and motivated a new generation of activists and leaders to get involved in the fight for affordable housing. In sum, HALA didn't end all of Seattle's problems, but it did outline what an ambitious pro-housing agenda looks like in a large, dynamic city.

Private entities have also played an important role in the affordable housing landscape in Seattle. Leading philanthropies such as the Bill & Melinda Gates Foundation, the Ballmer Group, and the Raikes Foundation have been actively involved in the fight against homelessness. Large corporations, including Amazon and Microsoft have provided billions of dollars to support the creation of affordable housing in the region. Amazon's investments alone now exceed $3 billion. Much of this money has been used to provide low interest loans that make financing affordable housing developments much easier. The significant response of the private sector speaks to the scale of the affordable housing challenges in Seattle.

Conclusion

Seattle is a city of contrasts. It is home to tremendous wealth and prosperity, but also has one of the highest rates of homelessness in the United States. How should we reconcile these seemingly inconsistent realities? The answer lies in the housing market. High housing costs are the result of regional prosperity, but they also exact a huge cost on low-income households. For the 1 in 10 Seattle households living below the poverty line, the $1,900 median rent would require 82% of income for a four-person household. For homeowners in Seattle, the multi-decade boom has been a source of tremendous personal prosperity as home equity serves a significant contributor to household wealth. For those households not fortunate to own a home, accessing housing has been increasingly difficult over the last two decades. Buying a home has become more difficult as home prices have risen dramatically, and finding and preserving rental housing is a challenge given the market forces that have driven rental costs upward. The reality of these challenges is evident in the increased focus on housing affordability and access by government, philanthropy, and corporations.

References

2009 Seattle Housing Levy Final Report. (2016). Office of Housing. Retrieved from www.seattle.gov/documents/Departments/Housing/Footer%20Pages/Levy-Annual-Report_2016.pdf

City of Seattle. (1987). *Urban refugees: The 1978 housing crisis – city archives*. Retrieved from www.seattle.gov/cityarchives/exhibits-and-education/online-exhibits/urban-refugees-the-1978-housing-crisis

City of Seattle. (2023). *Brief history of Seattle – city archives.* Retrieved from www.seattle.gov/cityarchives/seattle-facts/brief-history-of-seattle

City of Seattle. (2024a). *2023 multifamily tax exemption annual report*. City of Seattle Office of Housing. Retrieved from www.seattle.gov/documents/Departments/Housing/Reports/2023_MFTEAnnualReport.pdf

City of Seattle. (2024b). *2023 mandatory housing affordability and incentive zoning report*. City of Seattle Office of Housing. Retrieved from www.seattle.gov/documents/Departments/Housing/Reports/2023_MHA-IZAnnualReport.pdf

Colburn, G., & Aldern, C. P. (2022). *Homelessness is a housing problem: How structural factors explain U.S. patterns*. Oakland, CA: University of California Press.

City of Seattle. (2023). *Creating affordable housing with a linkage fee – council*. Retrieved from www.seattle.gov/council/issues/past-issues/creating-affordable-housing-with-a-linkage-fee

Durning, A. (2020). *One of North America's Boldest Housing Initiatives Has Reached Its End: Did It Work?* Sightline Institute. Retrieved from www.sightline.org/2020/02/21/one-of-north-americas-boldest-housing-initiatives-has-reached-its-end-did-it-work/

History | Seattle Housing Authority. (2023). Retrieved from www.seattlehousing.org/about-us/history

Housing Trust Fund (HTF). (2023). Washington State Department of Commerce. Retrieved from www.commerce.wa.gov/building-infrastructure/housing/housing-trust-fund/

Lange, G. (1999). Billboard reading "Will the Last Person Leaving SEATTLE – Turn Out the Lights" appears near Sea-Tac International Airport on April 16, 1971. HistoryLink.org. Retrieved from www.historylink.org/file/1287

Levy, D. K., Comey, J., & Padilla, S. (2007). In the face of gentrification: Case studies of local efforts to mitigate displacement. *Journal of Affordable Housing & Community Development Law*, *16*(3), 238–315.

Puget Sound Regional Council. (2022). *Regional Housing Needs Assessment*. Retrieved from www.psrc.org/sites/default/files/2022-09/rhna.pdf

San Telmo Associates v. City of Seattle. (1987). Justia Law. Retrieved from https://law.justia.com/cases/washington/supreme-court/1987/52978-7-1.html

Seattle Office of Housing. (2023). *A history of maximizing housing levy dollars*. Retrieved from https://seattle.gov/documents/Departments/Housing/Programs%20and%20Initiatives/Levy/Levy_Accomplishment.pdf

Seattle Voters Overwhelmingly Pass Housing Levy, Continue Legacy of Investment |. (2016). Retrieved from https://housingtrustfundproject.org/seattle-voters-overwhelmingly-pass-housing-levy-continue-legacy-of-investment/

State preempts certain tax fields – Fees prohibited for the development of land or buildings – Voluntary payments by developers authorized – Limitations – Exceptions, Revised Code of Washington § 82.02.020 (1982). Retrieved from https://app.leg.wa.gov/RCW/default.aspx?cite=82.02.020&pdf=true

U.S. Census Bureau QuickFacts: Seattle city, Washington. (2023). Retrieved from www.census.gov/quickfacts/fact/table/sanantoniocitytexas,seattlecitywashington/PST045222

Part V

The Path Forward

Chapter 15

An Affordable Housing Roadmap

Introduction

Throughout this book, we have sought to define what affordable housing is, describe its landscape, highlight the consequences of unaffordable housing, explain how affordable housing is created and preserved, and how different actors and institutions interact to produce this outcome. Regardless of one's political or ideological perspective, there is broad agreement that the cost of housing is a significant challenge for far too many households in the United States.

In this final chapter, we write from the perspective that the current unaffordability of housing in the nation is a significant problem that requires and demands a new approach. But outlining a path forward is challenging – ideal policies and programs are easy to describe, but harder to implement: some levels of government may be willing or able to make changes, while others are not; real or perceived budgetary constraints may restrict the ability to fund programs at scale; and people from different ideological perspectives may have very different ideas on how best to address this problem. In addition, existing public perceptions about affordable housing and low-income households can be impediments to broader progress.

In this final chapter, we chart a course for an improved housing system in the United States. To clarify and inform our roadmap, we draw on our own research, but also rely on the perspectives from the robust community of housing advocates, analysts, policymakers, and scholars. Because of the complexity of the task, we do not outline one path, because different factors could determine whether that outcome could be achieved. Rather, we highlight four different *states* (or outcomes) that could be achieved with varying levels of political will and resources. The first state considers changes that could be implemented in a status quo political and economic environment and the final state considers what transformational change in our nation's approach to housing might look like.

Housing is not considered a fundamental human right in the United States, and government assistance lacks the necessary scale to guarantee such a right. But the idea of housing as a basic right is not a fringe opinion in the United States. Eric Tars of the National Homelessness Law Center wrote, "Polling indicates that three-quarters of Americans believe that adequate housing is a human right, and two-thirds believe that government programs need to be expanded to ensure this right" (Tars, 2018). But to date, housing policy in the U.S. bears little relationship to the polling results cited by Tars. Therefore, to achieve significant change in our nation's approach to housing, there will need to be a shift in American perceptions, and more importantly, a shift in policymaking to bring about that result. The prevailing and dominant ideology around housing in the U.S. has been about homeownership and how owning one's home can be a source of wealth creation. That ethos has benefited millions of families and households throughout the nation's history, but has also left a significant percentage of households – who have not benefited from homeownership – on the outside looking in. The question, therefore, is whether this approach and ideology will persist or whether the nation will chart a different course.

Readers might appropriately wonder what factors or conditions determine what state we might ultimately achieve. To answer that question, we provide a list of Fundamental Questions that appeal to basic principles of pragmatism. We could propose a fundamental change in U.S. policymaking, but if such changes are politically impossible, it is simply a thought exercise – a utopian outcome that has little tie to reality. In her book, *Fixer Upper*, Jenny Schuetz highlights the important role that political coalitions play in driving meaningful change (Schuetz, 2022). Schuetz notes that new coalitions should include greater numbers of younger renters, and households of color, who have not been adequately represented in prior housing policy debates. The interests of the real estate industry and older single-family homeowners have been disproportionately represented in these debates throughout history. Meaningful progress will require additional voices to be represented – and heard – in policy advocacy that seeks greater housing access in the United States.

The answers to the following five Fundamental Questions will help determine which of the states that we present below are most likely. Note that, in addition to these questions, larger geopolitical events could impact the state in which we find ourselves. For example, an accelerating climate crisis or a major political upheaval could hasten change in a way that conventional political changes might not. In general, the states below are not mutually exclusive and distinctions between the states may blur. We create these states to help readers understand different contexts in which we may find ourselves. These Fundamental Questions are an attempt to help us understand what decisions and actions may drive us to a given state or outcome:

1. The limited social safety net in the United States increases housing precarity. Will the nation pursue a more robust system of cash assistance (a more robust safety net) that would produce – by extension – greater housing affordability and attainability?
2. Will the political landscape, and associated political will, provide support for changes to housing policy that are negligible, incremental, substantive, or transformational? A recent example of meaningful change appeared in the proposed Build Back Better legislation passed by the House of Representatives in 2021. The bill was never passed and the proposed investments in housing were not represented in subsequent legislation.
3. How will legislative action (or lack thereof) at the federal level influence policymaking at the state and local level? For example, will continued inaction at the federal level necessitate a stronger response from state and local governments? Alternatively, would a significant federal response provide resources and motivation for lower levels of government to make the fundamental changes that are necessary?
4. Does the rapidly growing crisis of affordable housing produce a disproportionate policy response? Does crisis create policymaking opportunities that would otherwise not exist? There are many examples in policymaking where new opportunities were pursued due to unique and challenging circumstances. Throughout the United States, we have seen an acceleration in housing-related policymaking as the crisis of unaffordable housing persists and worsens. The response to this growing crisis will have a significant impact on which of the states (described below) we find ourselves in.
5. At a local level, will there be greater space (politically and spatially) for alternative forms of land use that will provide for additional multi-family housing, or will single-family zoning remain the default approach to land use? Without significant changes in land use, meaningful progress toward housing affordability will be difficult.

To help highlight the different paths that we, as a nation, might pursue related to housing, we have conceptualized four different future states: Status Quo, Incremental Change, Substantive Change, and Transformational Change. Where we land among these different states is directly related to the five Fundamental Questions that we pose above. Getting to the Transformational Change State will not happen if our nation's response to the Fundamental Questions is status quo. Readers of this book may find some potential States more favorable or attractive than others. We are clear that States Three and Four provide more robust support and housing access for low-income households than do the first two. To outline these different states, we rely on a range of books, articles, and reports from trusted sources. These different outcomes reflect the realities of political pragmatism – without changes in politics, our approaches will likely not change in any material respect. While we believe that States Three and Four present the preferred paths to achieve greater access to housing for the greatest number of households in the United States, political realities and impediments may prevent achieving one of these outcomes.

It is important to consider the relationship between federal and local policymaking when considering the states presented below. In States One and Two, there is limited to modest change made at the federal level. This inaction places significant pressure on states and localities to respond with a more forceful policy response. In States Three and Four, the federal government plays a far more significant role in the provision of affordable housing thus releasing lower levels of government from some of these responsibilities. In these states, state and local governments are primarily focused on land use and regulatory matters while the vast majority of subsidy and support is provided by the federal government.

State One: Status Quo

In this first state, major policymaking that affects affordable housing doesn't materially change from where it currently stands. Such a state is likely if the ongoing political polarization – which is currently prevalent – continues to limit meaningful policy change. In this state, we see limited changes in federal housing policy. In this outcome, the Housing Choice Voucher program remains the primary mechanism through which we support low-income families, yet the program remains wholly inadequate, with only a small fraction of eligible households receiving support. On the production side, the Low Income Housing Tax Credit continues to support the production of affordable housing, but the demand for such tax credits will continue to far exceed their supply.

In this status quo environment, a key emphasis of affordable housing advocates is to *prevent* further cuts in already underfunded and limited housing supports. For example, in the National Low Income Housing Coalition's 2022 Public Policy Priorities, "protecting" and "preserving" existing programs is frequently mentioned. While the existing slate of programs and policies is inadequate, any steps backward would have devastating consequences, hence the NLIHC's focus on *preservation* and expansion of existing programs (National Low Income Housing Coalition, 2022).

In this state, housing affordability pressures will persist, and likely escalate, throughout the country. These challenges will place additional pressure on state and local governments to act. Despite the obvious needs, political impediments will complicate the policymaking efforts of states and localities. The knotty problem of local politics continues to impede zoning and land use changes that are needed to dramatically increase the housing supply in the United States. Fiscal limitations also constrain local policymaking which prevent states and local jurisdictions from filling the substantial, and well-documented, housing subsidy gap left by the federal government. The housing system remains highly fragmented, resulting in burdensome administration and an inefficient use of valuable, yet scarce, resources.

State One: Status Quo does not preclude meaningful progress. Even in the challenging political environment of the last few decades, advances have been achieved. For example, the State of Oregon passed a statewide rent regulation bill in 2019 that limits rent increases to 7% plus inflation – subject to a cap of 10% as of 2024 (National Low Income Housing Coalition, 2019; Buckley, 2023); the City of Minneapolis eliminated single-family zoning and made other substantive regulatory changes throughout the entire city (Minott, 2023; Blumgart, 2022; Liang et al., 2024); the City of Seattle passed a nearly one billion dollar housing levy that will fund new affordable housing development and provide rental assistance to low-income households (Housen, 2023; Oron, 2023); and Los Angeles passed Proposition HHH which provided over a billion dollars for housing and services for people experiencing homelessness (Local Housing Solutions, 2021; Galperin, 2022). While none of these programs was a silver bullet for the problem of affordable housing – and there are many well-documented issues and repercussions associated with these initiatives – their passage highlights that even in a constrained political environment, state and local governments can make meaningful progress toward improving access to housing. Even the federal government has made progress – albeit modest relative to the scale of the problem – as it established the National Housing Trust Fund in 2008 as part of the Housing and Economic Recovery Act of 2008. In 2023, $382 million was made available to states from the trust fund. As the National Low Income Housing Coalition notes on its website, the National Housing Trust Fund "is the first new housing resource since 1974 targeted to the building, rehabilitating, preserving, and operating rental housing for extremely low-income people" (Gramlich, 2022; Orbach, 2024). Admittedly, the federal trust fund was established during the 2007–2009 Great Financial Crisis so one could argue that this was not the result of "status quo" policymaking.

The lack of action from the federal government in State One places significant pressure on lower levels of government to provide policy responses. Earlier in this section, we listed a range of actions taken by states, counties, and cities to advance efforts to create affordable housing. We anticipate that such efforts will continue – and potentially accelerate – in State One (largely out of necessity). The list of potential actions is long, but could include state and local housing trust funds, rental assistance programs, tax credit programs to supplement LIHTC, focus on transit-oriented housing development, reducing single-family zoning, fast track permitting for certain types of affordable housing, rent regulation, and support for diverse modes of housing that might provide different living arrangements at lower prices.

Beyond the governmental responses highlighted above, in State One we anticipate ongoing activity by the private sector. In the last decade, we have seen greater involvement in affordable housing by a range of private actors throughout the country. Certainly, some of this activity comes from private non-profit and philanthropy that have long supported affordable housing in the communities that they serve. But recently, we have seen greater involvement from for-profit enterprises that recognize the risks posed by inadequate affordable housing. For example, in Seattle and the Puget Sound region, Amazon and Microsoft have made major commitments toward affordable housing. Such commitments are being made in cities throughout the United States. Interestingly, these multi-billion commitments have been made using corporate funds, not philanthropic dollars. This distinction is important because it indicates that these corporations have prioritized supporting affordable housing as a prudent use of corporate resources. Leaders of both companies have indicated that a lack of affordable housing is a risk for their organizations and the region as a whole. In response, Amazon's Housing Equity Fund is scheduled to make over $3.6 billion available for low interest loans and grants to support the creation of 35,000 units of affordable housing. In 2019, Microsoft made a corporate commitment toward affordable housing that now stands at $750 million. In State One, we would expect continued private sector activity in this space given the inadequate response of the federal government.

Finally, in State One we anticipate an increase in grassroots efforts – from a range of organizations – to find and create dedicated funding to grow the affordable housing supply. While grassroots efforts will be important in each State, they are of increasing importance in the status quo environment given the lack of government involvement. Initiatives such as community land trusts and alternative housing options, like cohousing, could be popular. Advocates will strive to raise funds for non-profit organizations, with a focus on securing dedicated funding sources for programs like state and local housing trust funds to expand and preserve the affordable housing stock. Non-profit and community organizations will work diligently to address the needs of the low-income households that are not adequately supported by government or private sector efforts.

State Two: Incremental Change

In the second state, we see modest changes at the federal level and continued progress from state and local governments. Potential changes observed in State Two would likely include modest increases to the Housing Choice Voucher program and Low Income Housing Tax Credit program by the federal government. We see evidence of this type of incremental change from multiple sources. First, the Biden Administration outlined a Housing Supply Action Plan in 2022. In addition, the Bipartisan Policy Center proposed The American Housing Act of 2023 that includes a range of housing related proposals that they believed had bipartisan support at the time of their proposal (Bipartisan Policy Center, 2023). These proposed changes, while meaningful, are not a radical departure from existing policymaking given the need for ratification in a highly divided policymaking environment.

In 2022, the Biden-Harris Administration released a Housing Supply Action Plan that outlined policies that could help create more housing – and particularly affordable housing – over the next five years. The elements of this plan included providing incentives to jurisdictions that reform their zoning and land use policies, creating new financing mechanisms to build and preserve more homes where financing gaps exist, providing more funding for housing (including an expansion of LIHTC), ensuring that more homes go to be people who will live in them (as opposed to investors), and working with the private sector to address supply chain bottlenecks that have limited the production of housing (The White House, 2022).

The American Housing Act proposal, published by the Bipartisan Policy Center (BPC), had three primary elements. First, it highlighted a range of policies that could be implemented to increase the supply of affordable housing in the United States. BPC proposed an increase in LIHTC allocations by 50%, a new tax credit to develop owner-occupied homes in distressed communities, incentives to eliminate harmful zoning and land use policies, and support for innovation in residential construction. Second, the proposal suggested policies that would help preserve the existing stock of affordable housing. This would be achieved by extending the affordability requirements of LIHTC and providing federal financing to upgrade or rehabilitate current affordable housing (including public housing). Finally, The American Housing Act proposed actions to help families afford housing. Included in this list of policy proposals were ideas to increase landlord participation in the voucher program and reduce discrimination in the housing market. What is conspicuously absent from this proposal from the BPC is an expansion of the voucher program. Its absence can likely be explained by the fact that voucher expansion did not enjoy bipartisan support at the time of this proposal. Therefore, both the Housing Supply Action Plan and the American Housing Act could make materially positive impacts on the housing system in the United States, but do not, as currently constructed, provide meaningful change in the housing policymaking landscape in the country.

What both of these proposals share is an effort to encourage certain behavior from local governments. In particular, to provide financial incentives for states and localities that alter land use and housing regulations to enable additional housing supply. These suggestions highlight that the federal government understands how important these local decisions are in determining how our nation can create the additional housing that is so desperately needed. Jenny Schuetz (2022) stresses the need to limit localism in housing policymaking. By giving cities, neighborhoods, and individual blocks such power over development decisions, localized preferences and biases can have a significantly negative impact on housing supply and, by extension, affordability. The proposals outlined above would provide incentives (from the federal government) that might help blunt the localism that has been a significant impediment to date.

State Two would also include a more robust defense of the principles of fair housing. As described more fully in Chapter 5, the constrained access to housing by households of color throughout U.S. history has had significant, and well-documented, consequences. In response, a host of fair housing regulations have been passed at multiple levels of government, starting with the Fair Housing Act of 1968. But incomplete and inconsistent enforcement of such laws has allowed unfair housing practices to persist. In State Two, we would anticipate greater emphasis on enforcement of the Fair Housing Act and other non-discrimination efforts like Affirmatively Furthering Fair Housing. The importance of these issues is evident in that a robust focus on fair housing is one of the key public policy priorities of the National Low Income Housing Coalition (National Low Income Housing Coalition, 2022).

State Three: Substantive Change

As we move to State Three and State Four, the major difference is the role of the federal government. We begin to see a much more forceful federal government response that could produce material change in the trajectory of affordable housing in the United States. The pillars of State Three include:

1 a demonstrable commitment to housing production,
2 a robust social safety net that provides more generous cash assistance and/or greater rental assistance to low-income households in the U.S.,
3 a robust suite of tenant supports at the state and local level that mitigates the power differential between landlords and tenants, and
4 eliminating government programs that support middle and higher income homeowners.

Shane Phillips, in his book, *The Affordable City*, identified his strategy for producing meaningful progress in affordable housing. He highlighted the Three Ss: Supply, Subsidy, and Security (Phillips, 2020). As you'll see, the first three pillars of State Three cover those three concepts. We need more housing, more subsidy for low-income households, and greater regulatory support for households that are precariously housed. Phillips' work is an excellent roadmap for the types of changes that would be needed to produce substantive improvement in the U.S. affordable housing system.

Recent efforts and pronouncements from the Biden Administration indicate policy directions consistent with the substantive change of State Three. First, the housing provisions included in the failed Build Back Better legislation would have provided an additional $170 billion in housing investments, including over 300,000 additional housing vouchers, $65 billion to address unmet maintenance needs in the existing public housing stock, and additional investments toward the national Housing Trust Fund ($15 billion), Low Income Housing Tax Credit ($12 billion), and HOME Investment Partnership ($10 billion). The elements of this proposal reflect what substantive changes in housing policy might look like (Fischer, 2022). More recently, President Biden, in his 2024 State of the Union Address outlined a plan to construct and rehabilitate two million homes. Initiatives of this scale are what is required to make meaningful change. Neither proposal became law, but the Build Back Better legislation was as close as the U.S. has been to major enhancements to housing policy in decades. In describing the proposals outlined in President Biden's State of the Union Address, Lael Brainard, Director of the National Economic Council, called it "the boldest housing plan in a generation" (Thomhave, 2024).

The four pillars of State Three (while not exhaustive) provide an opportunity for substantial change in our nation's approach to housing. First, there must be a demonstrable commitment to create more housing in the United States. These commitments should come from all corners of society – all levels of government, corporations, and public and nonprofit enterprises. Various estimates suggest a deficit of housing in the United States of between 3 to 7 million housing units. Therefore, societal commitments must total in the millions, not thousands, of units. The scale described by President Biden in the 2024 State of the Union Address is the level at which we need to be discussing housing production. Large goals need to be set and then multiple sectors of society must work together to figure out how to accomplish that goal. In State Three, the government wouldn't develop and manage the housing, but it would play a key role. First, governments at all levels could provide resources, either in the form of low-cost financing or by providing subsidy to help fund housing projects – particularly those that are affordable. The second role for government in State Three is to ensure that land use and regulation is aligned with the goal of creating more housing. More broadly, we also need to consider whether the housing production system is ready to build this scale of housing. Do we have a sufficient number of development firms to construct this much housing? Is there sufficient financing available? Do we have a sufficient labor force to construct this number of units? Do we have new construction technologies that will allow us to construct housing more quickly and more affordably? All sectors of society will need to come together to provide affirmative answers to these important questions.

The second pillar requires a more robust social safety net, such protections could be provided via more significant cash assistance or greater provision of rental assistance. There is growing evidence and support for the idea of giving people cash rather than structuring in-kind programs such as housing vouchers. One of the chapters in Jenny Schuetz's book, *Fixer Upper*, is titled "Give poor people money." Schuetz (2022) argues that the limited social safety net accentuates housing challenges for low-income households. Providing more financial resources to these households would be an important step in improving housing access for millions of Americans. Emerging data from a range of basic income experiments underscores the effectiveness of unrestricted cash payments to help promote affordable housing. In a summary of basic income evidence, researchers from Urban Institute found that cash-based assistance provides multiple benefits, including greater housing stability for renters, greater residential choice, reduced discrimination, improved housing access, and greater efficiency in housing delivery (Bogle et al., 2022). Casey Dawkins, in his book, *Just Housing*, argues for converting the Housing Choice

Voucher program into a cash assistance program (Dawkins, 2021) for many of the same arguments that are outlined above.

In the absence of creating a more robust system of cash assistance, another way to create a more robust safety net is to expand the current housing support programs that already exist. An obvious example would involve making the Housing Choice Voucher program an entitlement. Rather than serving only 1 in 4 eligible households – as it currently does – making the program an entitlement would ensure a housing voucher for everyone who qualifies. This would bring housing support into alignment with the way the country provides food support through the Supplemental Nutrition Assistance Program (SNAP) which is also an entitlement. President Biden proposed making the voucher program an entitlement during the 2020 Presidential campaign and many researchers have discussed the proposal and estimated its cost. Mary Cunningham of Urban Institute published a study entitled "It's Time to Reinforce the Housing Safety Net by Adopting Universal Vouchers for Low-Income Renters" in which she estimated that making the Housing Choice Voucher program an entitlement would cost $62 billion, based on analysis from 2020 (Cunningham, 2020). While $62 billion is certainly a significant sum, it would represent a modest percentage of the $1.7 trillion discretionary portion of the federal budget.

Pillar three would also involve a robust set of tenant protections that would seek to create a greater balance in power between landlords and tenants. Phillips (2020) highlights *security* as one of the keys to creating more affordable cities:

> stability is about recognizing the dignity of housing – that it's more than an investment vehicle and a means of creating personal wealth, as it is often treated today. It relates to tenant protections and rental housing preservation, two overarching programs that ensure all residents have safe, clean, affordable housing without fear that the rug might someday be pulled out from under them.
>
> (Phillips, 2020, p. 19)

Phillips provides a long list of approaches that contribute to greater tenant protection and less precarious housing for many low-income households. His list of proposals include: anti-gouging rental regulation, density bonuses to promote more (and dense) housing, inclusionary zoning, avoiding development that exacerbates displacement, preferred placement for displaced residents, just-cause eviction protections, free or reduced legal counsel to tenants facing eviction, and enforcing anti-discrimination provisions in the housing market. There is not one provision that will completely protect tenants, but there are a range of regulatory provisions that, in combination, can create far more housing security for low-income households.

Finally, the fourth pillar is a source of funding. As highlighted earlier in this book, there are a range of housing policies that disproportionately benefit middle- and upper-income households. The most obvious is the home mortgage interest tax deduction. While the lost revenue associated with this tax deduction fell after the 2017 tax reform legislation, it is still significant. According to estimates from the Brookings Institution, the post-tax reform cost of the deduction was about $30 billion in 2019 (Gale, 2019). This change alone would be sufficient to fund nearly half of the cost of making Housing Choice Vouchers an entitlement. This and other changes would also be an important signal to the country that the focus of the nation's housing policy will be on affordability rather than on enriching middle- and upper-income households by promoting and subsidizing homeownership.

While the changes outlined in State Three are substantive, they do not meaningfully alter the housing system in the United States. In this state, private market actors continue to construct

and operate the vast majority of housing in the nation. But the policy changes outli[n]
reduce the commodification of the housing system. Households become less reliant [on the pri]
vate market for their basic needs (including housing) because of a more robust so[cial safety]
net, housing assistance that is provided based on need (not rationed), and tenant regulations
that provide renters with additional power in the landlord/tenant relationship. In this State, the
private market still plays a fundamental role in the housing system, but its influence and power
are mitigated through policymaking.

The difference between State Three: Substantive Change, and the final state presented below, is that State Three could conceivably be achieved in the existing political environment. The ideas presented in State Three do not fundamentally change the free market system on which the U.S. economy is based and it leaves the private market as the dominant provider of housing in the country. Given current political constraints, one could certainly argue that the changes outlined in this third state are not feasible, but we do believe that this level of change is a realistic goal for policymakers throughout the country.

State Four: Transformational Change

In State Four, we relax the constraints of political feasibility and consider what a significant transformation of the U.S. housing system might look like. The most significant difference between the third and fourth states is the role of the government in the provision of housing. The bedrock of State Four would be a national articulation of housing as a basic human right. Other countries, such as the Netherlands, have enshrined a right to housing in their national constitution. If the U.S. were to join other nations in ensuring housing as a basic right of citizenship, it changes the role of the federal government meaningfully.

In such a case, the government would no longer delegate housing production almost entirely to the private sector. Rather, the government would, once again, become an active participant in the housing system through government sponsored public or social housing. A robust public/social housing program would serve as the basis of State Four. Rather than using tax credits and vouchers to encourage private market actors to provide affordable housing, in State Four the government would finance, build, and operate housing. A review of OECD countries shows that Netherlands (34%), Austria (24%), Denmark (22%), United Kingdom (16%), France (14%), Ireland (11%), and Finland (11%) all have social housing stocks that represent greater than 10% of the total housing stock in their respective countries (OECD, 2024). As a reminder, the equivalent figure in the United States is less than 1%. This is why we describe this state as Transformational Change because the current commitment to public or social housing in the U.S. is de minimis. Of course, a larger stock of public housing would not alter the fact that the private market would still construct and operate the vast majority of housing units in the country, but the public sector would play a far more significant role that it currently does.

Success in State Four goes beyond simply having the government construct and operate more housing, although that is an important part of this outcome. State Four would also involve a strong commitment to tenants' rights and rental regulation as outlined in State Three. The difference would be that rather than the government attempting to regulate the almost entirely privately-owned stock of rental housing, a larger percentage of the housing stock would be under the government's control and therefore easier to enforce the rules and regulations that it has established. The government could also make an effort to ensure that new, government-sponsored, housing would be located in a range of different neighborhoods. In the past, public housing was largely located in high poverty and under-resourced neighborhoods, but new commitments to publicly-owned housing could provide greater access to a diverse set of neighborhoods for residents of this housing.

One of the key challenges of State Four is to determine how this public housing would be created or acquired. First, whether the government or private developers construct the housing, the United States needs to create a more efficient system of housing delivery. We currently face constraints in the development of housing both financially and operationally. Even in a state where the government owns and operates a larger share of the housing stock, there will still need to be a robust supply of private firms that are able to procure the necessary supplies and construct the housing. Regardless of the state in which we find ourselves, we must have a national conversation about ensuring that we have sufficient supplies, materials, and labor to address our deficit of housing. In 2022, the United States passed the CHIPS and Science Act to prioritize the importance of the semiconductor industry in the nation. Having control over our own semiconductor supply chain became an issue of national security and lack of control jeopardized the economic viability of the country. In a similar way, the government needs to recognize that our nation's inability to construct adequate housing also has a significant impact on the health and wellbeing of the country.

A major difference between State Four and the other potential outcomes is that housing is less integrated into the financial system. Government doesn't use its power to support the private market and middle- and higher-income homeowners don't receive significant tax benefits from the government. While not eliminating the idea that housing is an asset and a source of wealth creation, State Four would shift the U.S. housing system toward an increased focus on the use value of housing rather than the asset value of housing. This is the essence of the idea of decommodification. When households are less reliant on the private market for the goods and services that they need and desire, their existence becomes more decommodified. Currently, the United States has a highly commodified economic system in which almost all households have significant exposure to (and dependence on) the private market. This system benefits some households, while it harms others. State Four would create a housing system in which low-income households are less dependent on the forces of the private market to procure the housing that they need.

There are existing examples and proposals that could provide a roadmap for what some of the policy initiatives in State Four might look like. Greater public ownership of housing does not necessarily mean that all of that housing needs to be financed and constructed by the federal government. The King County Housing Authority (KCHA) in Washington has been an active acquirer of multi-family housing over the last two decades. As of October 2022, KCHA owned nearly 12,500 units of housing. One of the motivations of these acquisitions was to take units out of the private market such that they are no longer subject to market forces. By acquiring these units, KCHA has contributed to a more decommodified housing stock and they are now in control of the rents for these units for as long as they own them (and can afford ongoing maintenance which has become a challenge for some public housing authorities). This is a highly effective way to ensure long-term affordability. Solely using vouchers to create affordability in a highly commodified housing system is challenging because as market rents go up, the cost of each voucher increases. There is no cap on the cost of vouchers when tenants are subject to market rents. KCHA has shown how an aggressive acquisition program can produce a portfolio of publicly-owned and operated housing that can deliver affordability for decades to come.

In 2023, voters in the City of Seattle approved Initiative Measure 135 which called for the city to develop, own, and manage a social housing program. The passage of the measure provided startup funding for the program, but not funding to actually acquire or operate the housing.

Those efforts remain ongoing. Supporters for the Seattle social housing initiative cited the efforts of Montgomery County, Maryland as an inspiration. Montgomery County has long been held up as a success story of a jurisdiction that has created a stock of publicly-owned housing in response to a growing housing affordability crisis. Conor Dougherty covered this effort in the *New York Times* in the article, "This is public housing. just don't call it that" (Dougherty, 2023). The title of the article refers to an inherent conflict in efforts of this nature. On one hand, there is the nation's unease with the idea of public housing – in light of our nation's history with these programs – and on the other hand, it acknowledges the important role that publicly-owned housing can play in providing and ensuring affordability in rapidly escalating housing markets. In Montgomery County, the local government has begun playing an important – yet still modest – role as a funder and owner of affordable housing while partnering with private developers to construct the housing. Dougherty describes the forces that motivated this local action:

> In Montgomery County, however, the stock of government-owned housing has steadily grown for decades while the definition of what it can be has expanded. The reason: In the Washington region, as in every other high-growth metropolitan area, the demand for affordable housing is way beyond what federal housing programs can provide. So the county tries to make up the gap.

Dougherty interviewed Andrew Friedson who is a member of the Montgomery County Council that approved and supported the housing fund that helped to develop this new publicly-owned housing. Friedson told Dougherty,

> We have to get out of the view that certain things are dirty words: "Public housing" is not a dirty word. "Developer" is not a dirty word. The market is not functioning the way we need it to, and that's when we want the government to step up.
>
> (Dougherty, 2023)

Even in State Four, in which housing becomes a right and the government plays a far more significant role, the public sector will continue to partner with private firms to construct and develop affordable housing. What changes is who controls the units and the primary motivation of the development. Rather than constructing housing to deliver a profit to investors, this form of housing would be constructed with the stated goal of delivering affordability in perpetuity.

Conclusion

By outlining the Four States in this chapter, we seek to highlight how different decisions could have radically different implications for the affordable housing system in the United States. Many policy books conclude with a chapter that describes the ideal world or outcome, but that potential world may not be grounded in reality. In a nod toward political pragmatism, we used the Four States to highlight what might be accomplished in a range of different political and economic contexts. By describing the Status Quo state, we are not excusing or condoning that outcome; we firmly believe that a status quo trajectory would be harmful for the nation and, in particular, the millions of low-income households that struggle to find and maintain housing. But we hope to highlight that the Status Quo state does not preclude progress *and* that there are other outcomes (States Two through Four) that are possible. Finally, these states are creations of our imaginations. They are not discrete or mutually exclusive; the ideas and policies described in each state could easily blur across different states. Rather than a firm prescription for future trajectories, we use the Four States as a thought exercise to highlight a variety of different potential outcomes.

Throughout this book we have attempted to provide readers with an introduction to the complex topic of affordable housing. Without a doubt, this effort is inadequate. Each of the chapters and concepts in this book deserved more attention. We attempted to strike a balance between being comprehensive, yet still accessible. Our hope is that readers who are new to this topic will have a better sense for the field, why it matters, the existing policy framework, and what new pathways forward might look like. We encourage readers with a deeper interest in the topic to use the bibliography in this book to dive more deeply into the outstanding scholarship that exists in this field. We view this book as the start, not the end, of your journey to understand the field of affordable housing.

In addition to the works cited elsewhere in this book, the following books, chapters, and articles would be excellent resources for readers looking to dive more deeply into these topics:

Anacker, K. (2024). *Housing in the United States: The basics*. New York: Routledge.
Been, V., Ellen, I. G., & O'Regan, K. M. (2023). Supply skepticism revisited. NYU Law and Economics Research Paper No. 24–12. https://papers.ssrn.com/sol3/papers.cfm?abstract_id=4629628
Clapham, D. F., Clark, W. A. V., & Gibb, K. (Eds.) (2012). *The Sage handbook of housing studies*. Thousand Oaks, CA: Sage Publications, Ltd.
Collinson, R., Ellen, I. G., & Ludwig, J. (2016). Low-income housing policy. In R. A. Moffitt (Ed.), *Economics of Means-Tested Transfer Programs in the United States, Volume II &* (pp. 59–126). University of Chicago Press.
Dawkins, C. J. (2021). *Just housing: The moral foundations of American housing policy*. Cambridge, MA: The MIT Press.
Freeman, L., & Schuetz, J. (2017). Producing affordable housing in rising markets: What works? *Cityscape, 19*(1), 217–236.
Galster, G., & Lee, K. O. (2021). Housing affordability: A framing, synthesis of research and policy, and future directions. *International Journal of Urban Sciences, 25*(0), 7–58.
Goetz, E. (2013). *New Deal ruins: Race, economic justice, and public housing policy*. Ithaca, NY: Cornell University Press.
Jargowsky, P. A., Ding, L., & Fletcher, N. (2019). The Fair Housing Act at 50: Successes, failures, and future directions. *Housing Policy Debate, 29*(5), 694–703.
Kneebone, E., & Reid, C. K. (2021). The complexity of financing low-income housing tax credit housing in the United States. UC Berkeley Terner Center for Housing Innovation.
Landis, J. D., & McClure, K. (2010). Rethinking federal housing policy. *Journal of the American Planning Association, 76*(3), 319–348.
Manville, M., Monkkonen, & Lens, M. (2019). It's time to end single-family zoning. *Journal of the American Planning Association, 86*(1), 106–112.
Marohn, Jr., C. L., & Herriges, D. (2024). *Escaping the housing trap: The strong towns response to the housing crisis*. Hoboken, NJ: Wiley.
Marcuse, P., & Madden, D. (2016). *In defense of housing: The politics of crisis*. London: Verso Books.
McCabe, B., & Rosen, E. (Eds) (2023). *The sociology of housing: How homes shape our lives*. Chicago: The University of Chicago Press.
Mueller, E., & Tighe, J. R. (Eds) (2012). *The affordable housing reader*. London: Routledge.
Phillips, S. (2020). *The affordable city: Strategies for putting housing within reach (and keeping it there)*. Washington, DC: Island Press.
Scanlon, K., Whitehead, C., & Arrigoitia, M. F. (Eds) (2014). *Social housing in Europe*. West Sussex: John Wiley & Sons, Ltd.
Schuetz, J. (2022). *Fixer upper: How to repair America's broken housing systems*. Washington, DC: Brookings Institution Press.
Schwartz, A. (2021). *Housing policy in the United States*, 4th edition. New York: Routledge.
Squires, G. (2017). Mechanisms for financing affordable housing development. In *Routledge Companion to Real Estate Development* (pp. 207–217). New York: Routledge.

References

Been, V., Ellen, I. G., & O'Regan, K. M. (2023). Supply skepticism revisited. NYU Law and Economics Research Paper No. 24–12. Retrieved from https://papers.ssrn.com/sol3/papers.cfm?abstract_id=4629628

Bipartisan Policy Center. (2023). *A bold, bipartisan response to the housing affordability crisis: The American housing act of 2023*. Washington, DC. Retrieved from https://bpcaction.org/wp-content/uploads/BPC-Action-Housing-Legislative-Brief.pdf

Bogle, M., Williams, J. L., Braswell, C., & Fung, L. (2022). *Guaranteed income as a mechanism for promoting housing stability*. Urban Institute. Retrieved from www.urban.org/research/publication/guaranteed-income-mechanism-promoting-housing-stability

Blumgart, J. (2022, May 26). How important was the single-family zoning ban in Minneapolis? *Governing*. Retrieved from www.governing.com/community/how-important-was-the-single-family-housing-ban-in-minneapolis

Buckley, K. (2023, September 26). New rent cap kicks in, limiting hikes to 10% next year for some Oregonians. *OPB*. Retrieved from www.opb.org/article/2023/09/26/oregon-rent-increase-caps/

Cunningham, M. K. (2020). *It's time to reinforce the housing safety net by adopting universal vouchers for low-income renters*. Urban Institute. Retrieved from www.urban.org/urban-wire/its-time-reinforce-housing-safety-net-adopting-universal-vouchers-low-income-renters

Dawkins, C. J. (2021). *Just housing: The moral foundations of American housing policy*. Cambridge, MA: The MIT Press.

Dougherty, C. (2023, August 25). This is public housing. Just don't call it that. *The New York Times*. Retrieved from www.nytimes.com/2023/08/25/business/affordable-housing-montgomery-county.html

Fischer, W. (2022). *Housing investments in build back better would address pressing unmet needs*. Center on Budget and Policy Priorities. Retrieved from www.cbpp.org/research/housing/housing-investments-in-build-back-better-would-address-pressing-unmet-needs

Freeman, L., & Schuetz, J. (2017). Producing affordable housing in rising markets: What works? *Cityscape*, *19*(1), 217–236.

Gale, W. G. (2019, May 13). *Chipping away at the mortgage deduction*. Brookings. Retrieved from www.brookings.edu/articles/chipping-away-at-the-mortgage-deduction/

Galperin, R. (2022). *The problems and progress of prop. HHH*. City of Los Angeles, Office of the Controller. Retrieved from https://controller.lacity.gov/audits/problems-and-progress-of-prop-hhh

Galster, G., & Lee, K. O. (2021). Housing affordability: A framing, synthesis of research and policy, and future directions. *International Journal of Urban Sciences*, *25*(0), 7–58.

Girten, N. (2024, May 10). Montana task force tackles adding more affordable housing. *Missoula Current*. Retrieved from https://missoulacurrent.com/montana-affordable-housing-4/

Gramlich, E. (2022). *The national housing trust fund: A summary of 2018 state projects*. The National Low Income Housing Coalition.

Housen, J. (2023). Mayor Harrell celebrates voters' historic passage of $970 million housing levy. City of Seattle, Office of the Mayor. Retrieved from https://harrell.seattle.gov/2023/11/07/mayor-harrell-celebrates-voters-historic-passage-of-970-million-housing-levy/

Jargowsky, P. A., Ding, L., & Fletcher, N. (2019). The Fair Housing Act at 50: Successes, failures, and future directions. *Housing Policy Debate*, *29*(5), 694–703.

Kneebone, E., & Reid, C. K. (2021). *The complexity of financing low-income housing tax credit housing in the United States*. UC Berkeley Terner Center for Housing Innovation.

Landis, J. D., & McClure, K. (2010). Rethinking federal housing policy. *Journal of the American Planning Association*, *76*(3), 319–348.

Liang, L., Staveski, A., & Horowitz, A. (2024, January 4). Minneapolis land use reforms offer a blueprint for housing affordability. Pew Charitable Trust. Retrieved from www.pewtrusts.org/en/research-and-analysis/articles/2024/01/04/minneapolis-land-use-reforms-offer-a-blueprint-for-housing-affordability

Local Housing Solutions. (2021, April 16). *Los Angeles Proposition HHH*. Retrieved from https://localhousingsolutions.org/housing-policy-case-studies/los-angeles-proposition-hhh/

Manville, M., Monkkonen, & Lens, M. (2019). It's time to end single-family zoning. *Journal of the American Planning Association*, *86*(1), 106–112.

Minott, O. (2023). *Comprehensive zoning reform in Minneapolis, MN*. Bipartisan Policy Center, Washington, DC. Retrieved from https://bipartisanpolicy.org/blog/comprehensive-zoning-reform-in-minneapolis-mn/

National Low Income Housing Coalition. (2019, March 4). *From the field: Oregon passes nation's first statewide rent control law*. Retrieved from https://nlihc.org/resource/field-oregon-passes-nations-first-statewide-rent-control-law

National Low Income Housing Coalition. (2022). *2022 public policy priorities*. Washington, DC. Retrieved from https://nlihc.org/sites/default/files/2022_Policy-Priorities.pdf

National Low Income Housing Coalition. (2023). *LIHTC Reform: Expansion must serve households with the greatest needs*. Washington, DC. Retrieved from https://nlihc.org/sites/default/files/LIHTC-Reform-Expansion-Must-Serve-Households-with-the-Greatest-Needs.pdf

OECD. (2024). PH4.2 Social Rental Housing Stock. OECD Affordable Housing Database. Retrieved from www.oecd.org/els/family/PH4–2-Social-rental-housing-stock.pdf

Orbach, R. (2024). *The National Housing Trust Fund and its impact to date*. Bipartisan Policy Center, Washington, DC. Retrieved from https://bipartisanpolicy.org/explainer/the-national-housing-trust-fund-and-its-impact-to-date/

Oron, G. (2023, October 11). What is the Seattle housing levy? Learn more about the initiative to fund affordable housing. *Real Change*. Retrieved from www.realchangenews.org/news/2023/10/11/what-seattle-housing-levy-learn-more-about-initiative-fund-affordable-housing

Phillips, S. (2020). *The affordable city: Strategies for putting housing within reach (and keeping it there)*. Washington, DC: Island Press.

Schuetz, J. (2022). *Fixer upper: How to repair America's broken housing systems*. Washington, DC: Brookings Institution Press.

Tars, E. (2018). *Housing as a human right*. National Low Income Housing Coalition. Retrieved from https://nlihc.org/sites/default/files/AG-2018/Ch01-S06_Housing-Human-Right_2018.pdf

The White House. (2022, May 16). President Biden announces new actions to ease the burden of housing costs. Retrieved from www.whitehouse.gov/briefing-room/statements-releases/2022/05/16/president-biden-announces-new-actions-to-ease-the-burden-of-housing-costs/

Thomhave, K. (2024, April 10). Biden has proposed 'the boldest housing plan in a generation,' his economic advisor says. *Multifamily Dive*. Retrieved from www.multifamilydive.com/news/biden-proposed-boldest-housing-plan-generation-economic-advisor-lael-brainard/712852/

Index

Page numbers in *italics* refer to figures and those in **bold** to tables.

30:40 Rule 7

ACCs *see* Annual Contributions Contracts
addiction 76
adjustable-rate mortgage (ARM) 67
affirmatively furthering fair housing (AFFH) 142–143
Affordable Requirements Ordinance (ARO) 167–168
AIG 66
air quality 62
almshouses 44
Amazon 189, 196–197
American Dream 18, 19, 40, 44, 47
American Housing Act of 2023 197–198
American Housing Survey 19
American Individualism 46
American Rescue Plan 69
AMI *see* Area Median Income
Annual Contributions Contracts (ACCs) 89, 96
Annual Homelessness Assessment Report (AHAR) 74
anti-discrimination laws 142
anxiety 61, 62, 68, 71
apartment buildings 16, 18, 162, 184
Apartments.com 61
Area Median Income (AMI) 9, 10, 101, 106, 114
Arizona 67
ARM *see* adjustable-rate mortgage
Asian households 54
Assets and the Poor New American Welfare Policy 130
Assets for Independence (AFI) program 130
asthma 61
Atkinson index 57
Atlanta 58
Australia 22
Austria 201

basic income 119–120
Bass, Karen 149–150
Bauer Wurster, Catherine 41
Bear Stearns 66
behavioral health 73
Below Market Interest Rate program 49
below market rate programs 49, 136
Bernanke, Ben 69
Better Homes in America program 45, 46
Biden, Joe 143, 197–200
Bipartisan Policy Center (BPC) 197–198
Black households 17, 40, 48, 54, 127; deed restrictions 59; eviction rate 55; homeownership rates 126; and residential segregation 55–56, 59
Black neighborhoods 59–61
Black people: and homelessness 55; and hypersegregation 57; and predatory loans 60; *see also* Black veterans; Black women; people of color
Black veterans 60
Black women 58
blockbusting 58–59
block grants 104–105; *see also* Community Development Block Grant (CDBG); HOME Investment Partnerships Program
Boston 73, 74
Breathe program 120
BRIDGE Housing Corporation 91
Brooke Amendment *see* Housing and Urban Development Act of 1969
Build Back Better legislation 199
building codes 21, 46, 88, 125, 149
building regulations 19, 23

Cabrini Green project 95, 98, 112, 160
California 8, 27, 58, 71, 88, 105, 108, 147
Canada 22, 37
Cannato, Vincent J. 45

capitalism 38
CARES Act of 2020 69
case studies: Chicago 159–168; San Antonio 169–177; Seattle 179–189
cash-based assistance 112, 117–122, 177, 194, 199–200
CBDOs see community-based development organizations
CDBG see Community Development Block Grant
CDCs see community development corporations
centralization 57
Chicago 9, 59, 95, 98, 117; federal housing supports 165; Gautreaux Project 116, 160; housing stock and market 161; overview 159–161; state and local housing supports 166–168
Chicago Community Housing Trust 167
Chicago Community Land Trust (CCLT) 167
Chicago Housing Authority (CHA) 160
Chicago Low-Income Housing Trust Fund (CLIHTF) 167
Chicago Neighborhood Rebuild Pilot Program 167
children: health outcomes 5; and homelessness 75, 76; from low-income households 62; and poverty 116–117; in public housing 96, 116–117; and Temporary Assistance for Needy Families (TANF) 118; see also infant mortality
Choice Neighborhoods 99
churches 42, 44
circuit breaker method 138
cities: ability to meet housing needs 41; coastal 73–75; housing cost burden across 27–30; population growth 35; residential segregation in 56
civic responsibility 45
Civil Rights Act of 1968 48, 141; see also Fair Housing Act
class theory 55
Cleveland 77
Clinton, Bill 118
closing costs 60, 86, 128–131
clustering 57, 58
co-housing 91, 92
Colonial Era 43–44
Colorado 88
commodification 37–38; see also decommodification
community-based development organizations (CBDOs) 91
community-based initiatives 91
community benefit agreements (CBAs) 150–151
Community Development Block Grant (CDBG) 84, 87, 91, 96, 104–105

community development corporations (CDCs) 42, 49–50, 91
community land trusts (CLTs) 50, 91, 135–136
community redevelopment agencies (CRAs) 108
Community Reinvestment Act of 1977 (CRA) 143–144
concentration, measuring 57
condos 132–133
Connecticut 147
cooperatives 134–135; limited equity 50; shared equity 136
couch surfing 72
COVID-19 pandemic 10, 69, 71, 73
CRA see Community Reinvestment Act of 1977
Craigslist 60–61
CRAs see community redevelopment agencies
crime 61, 68
criminal records 73
Crowder, Kyle 57

Dallas 103
Darden, J. T. 55
Dawkins, Casey 199–200
Dayton 29
decentralization 170
decommodification 38, 50, 202
deed restrictions 58–59, 136
Deficit Reduction Act of 1984 132
Denmark 37, 201
Denver 120
depression 61, 68, 71, 76
desegregation 116
Desmond, Matthew 70
detached homes 15, 16, 17, 145
Detroit 8, 26, 27, 77
disabled people 96, 113, 114, 119
discrimination: in housing market 16–17, 58–61, 142; location-based 58–59; protection from 141–142; see also discriminatory practices; racial discrimination
discriminatory practices 58–61, 103–104
disinvestment 39, 62–63, 96, 103
disparate impact 103–104
displacement 4, 48, 62
doubling up 21, 72
down payment assistance 129
Downpayment Toward Equity Act 129, 130

Earned Income Tax Credit (EITC) 118–120
Eastern European countries 18
economic disadvantage 17
economic inequality 57, 74
economic mobility 44
educational access 62
educational attainment 60
educational outcomes 54–55

EITC *see* Earned Income Tax Credit
Enhanced-Use Lease (EUL) program 87
Enterprise Community Partners 50, 90
entropy index 57
environmental hazards 62
Esping-Anderson, Gosta 38
evenness, measuring 57
Eviction Innovation project 71
evictions 69–72; consequences of 71–72; constructive 71; informal 71; just cause 71; and people of color 54, 55; prevalence of 70–71; types of 71; without cause 71
expedited permitting 149–150
Experimental Housing Allowance Program (EHAP) 113
exposure, measuring 57
extremely low-income households 9, 21, 102, 114, 117, 167, 173, 177

Facebook 151
Fair Housing Act of 1968 48, 56, 104, 141–143, 198
faith-based organizations 90
family 44; *see also* single-family homes; multi-family housing
Fannie Mae *see* Federal National Mortgage Association
Federal Deposit Insurance Corporation (FDIC) 143
federal government: encouragement of homeownership 44–46; historical perspective 43–51; housing programs and policies 5–6, 21–22, 38–43, 56, 96–105, 165–166, 173–174, 186; regulatory strategies 141–144; role of 46, 83; and subprime mortgage crisis 68–69
Federal Home Loan Bank System 84, 85, 86
Federal Home Loan Mortgage Corporation (Freddie Mac) 69, 85
Federal Housing Administration (FHA) 17, 47, 59–60, 131
federal housing development programs 96–102
Federal Housing Finance Agency (FHFA) 85–86
Federal National Mortgage Association (Fannie Mae) 69, 84, 85
Federal Reserve Bank 69
FHA *see* Federal Housing Administration
FHFA *see* Federal Housing Finance Agency
"filtering" 23
Finland 201
First-Time Homebuyer Act 132
first-time homebuyers 128–129, 131–132
Fixer Upper 194, 199
Florida 27, 67, 108
food insecurity 68
forbearance 69

foreclosure 46, 60, 66–69; consequences of 68–69; definition 66; judicial 67; power of sale 67; prevalence of 67, *68*; prevention initiatives 138; statutory 67; strict 67; types of proceedings 67
Fort Worth 61
foundations 91, 188–189
France 201
Freddie Mac *see* Federal Home Loan Mortgage Corporation
From the Puritans to the Projects 37
Furman Center (New York University) 90

Gautreaux, Dorothy 160
Gautreaux Project 116, 160
gender 44, 141, 142
gentrification 62, 85, 137
Germany 18
GI Bill 60
Gini coefficient 57
Gini index 58
global financial crisis 66, 86, 131, 196
global financial system 66
Goetz, Edward 49
GoSection8 (online platform) 61
government agencies 84–90
government-sponsored enterprises (GSEs) 85–86
grassroots agencies 91, 175, 197
Great Depression 39, 40, 44–47
Great Recession 15, 20, 138
GSEs *see* government-sponsored enterprises
gun violence 61

Habitat for Humanity 90
Hawaii 27
HCV program *see* Housing Choice Voucher (HCV) program
healthcare 62
heart disease 61
heating 19, 97
HFAs *see* state housing finance agencies
higher-income households 7, 10, 17–18, 23, 39, 51, 62, 98, 126–127
Hispanic households 40, 54, 127
HOLC *see* Home Owner's Loan Corporation
home equity 17, 32, 40, 126
HOME Investment Partnerships Program 87, 91, 96, 104–105, 174, 199
homelessness 4, 15, 21, 72–77; causes of 73–74; chronic 75–76; consequences of 76; cost of 76; definition 72; and Great Depression 46; interventions 76–77, 85, 86; and people of color 54, 55, 73, 75; and poverty 73–74; prevalence of 72–73; sheltered 74–75; types of 74–75; unsheltered 74–75
Homelessness is a Housing Problem: How Structural Factors Explain U.S. Patterns 183

210 Index

homeownership 10, 39–41, 125–139; affordability measures 6–7; assistance programs 129–132; ethos of 40; exclusionary practices in 58–60; and household income 17–18, 35; and housing cost burden 27–31; and national identity 46; and people of color 17, 40, 54–56, 125–126; preference for 16, 18, 40; preserving 137–139; promotion of 39, 40, 41, 45, 126; rates of 15, *16*, 17, 18, 40, 69, 125–126; shared equity 135; suburban 45, 47; and wealth generation 17–18, 31, 40; and wealth inequality 17, 40, 54
Home Owner's Loan Corporation (HOLC) 46–47
home prices 40, 54–55, 66, 109, 128, 137, 189; *see also* home values; property values
home repair programs 137
Homestead Act of 1862 44
home values: decline of 66, 67, 68; and wealth creation 18, 31, 127–128; *see also* home prices; property values
Hoover, Herbert 45–46
"Hoovervilles" 46
HOPE VI program 98–99
house boats 15
household budget 4, 6, 34
household composition 54, 73
household expenditures 5, 7, 118, 119
household income: and housing costs 4–7, 26–27, 34–35; and housing tenure 17–18, 34–35; measures 9; racial disparities in 54; *see also* higher-income households; lowest-income households; low-income households; middle-income households; moderate-income households; upper-income households
household size 18, 20
households of color 16–17, 34, 39, 61–62; *see also* Black households
house prices *see* home prices
Housing Act of 1934 47
Housing Act of 1937 41, 47, 89
Housing Act of 1949 39, 47, 48, 51
Housing Act of 1959 101
housing adequacy 19, 20, 23
housing affordability: achieving 9–10; crisis 3–5; definition 3; measuring 5–8; threshold for 6, 7; *see also* housing affordability roadmap
Housing Affordability and Livability Agenda (HALA) 188
housing affordability roadmap 193–203; Fundamental Questions 194; Incremental Change 197–198; Status Quo 195–197; Substantive Change 198–201; Transformative Change 201–203
Housing Affordability Stress 7
Housing and Community Development Act of 1974 48, 49

Housing and Economic Recovery Act of 2008 85, 102, 131, 196
Housing and Urban Development Act of 1969 (Brooke Amendment) 6, 48, 97
housing assistance 22, 26, 56, 61, 83, 84, 89–90; demand-side 10, 112–122; federal 173–174; local 22, 174–176; state 22, 174–176; supply side 95–110; *see also* housing support; public assistance
housing bonds 96, 106–107
Housing Choice Voucher (HCV) program 5–6, 21, 38, 49, 50, 61, 76, 83, 86, 87, 89, 104, 112–117, 120, 160, 174, 199–200; recipients 113–114
housing construction 10, 19–23, 37; cost of 22–23; and economic strength 46; encouraging of 47; innovation in 22, 23
housing cost burden 7, 8, 26–36; changes over time 30–34; across cities 27–30; median level 29, **30**, 31; and people of color 54
housing costs: and homelessness 74; and household income 4–7, 26–27, 34–35; and housing affordability 9, 10, 26; inflation rate for 35; *see also* housing cost burden
housing density 15; *see also* residential density; zoning
Housing First 76
housing instability 66–77, 138
housing market: Chicago 161–165; collapse 66; discrimination in 16–17, 58–61, 142; "filtering" 23; interaction with labor market 8–9; Seattle 181–185; *see also* private market
Housing Partnership Network 50
housing policy 5–6, 21–22, 38–51, 56, 85; context for 38–39; current landscape 39–43; discrimination in 39–40; encouraging homeownership 39, 40, 41, 45; historical development 43–51; racial disparities 39; two-tiered system 39, 41, 43; *see also* federal government; local governments; public policy; state governments
housing production 4, 9, 21, 45, 49, 58, 102, 147–148; *see also* housing construction; housing provision
housing provision 37–38, 88, 199
housing quality 9, 18–21
housing search process 57
housing shortage 20–23, 45, 47, 108–109, 185, 188; *see also* housing adequacy
housing size 19–20
housing starts 109
housing stock 15–24; average age 19; Chicago 161–165; publicly-owned 22; quality 9, 18–21; San Antonio 171–173; Seattle 181–185; typical lifespan 19
housing supply 20–22, 35, 97, 100, 108, 146, 154; *see also* housing assistance

Housing Supply Action Plan 197–198
housing support 42, 77, 120–121, 200; demand-side 112–113; federal 21–22, 38, 41, 87, 165–166, 173–176, 188; limited nature 21–22, 41, 114, 195; responsibility for 85; state and local 87–89, 166–168, 186–189; supply-side 95; *see also* housing assistance
housing systems 37–38
housing tenure 15–18; definition 15; and household income 17–18, 34–35; international comparisons 18; and race 17; types of 15–17
housing trust funds 43, 51, 87, 96, 105–106, 174–176, 196–197; *see also* Chicago Low-Income Housing Trust Fund (CLIHTF); National Housing Trust Fund
housing types 15–18, 132–133, 136
housing wage 7
Houston 146
HUD *see* U. S. Department of Housing and Urban Development
human rights 193, 201
hypersegregation 57

ICE *see* Index of Concentration at the Extremes
"ideology of property" 16
Illinois 166–168
Illinois Housing Development Authority (IHDA) 167
immigrants 44, 45, 90, 169
impact fees 150
inclusionary housing 147–148; *see also* inclusionary zoning
inclusionary zoning 10, 88, 147–148
Inclusive Communities 103
income inequality 34, 56
income transfers 118–122
independence 40
Index of Concentration at the Extremes (ICE) 58
index of dissimilarity 57
Indiana 27
Indian Housing Block Grant (IHBG) 105
Indigenous households 54
individual development accounts 130
individualism 44
Industrial Revolution 44
infant mortality 58
infill housing 148
interest rates 35, 67
Ireland 201

Japan 18, 37
Jim Crow laws 60
Joint Center for Housing Studies (Harvard University) 90

Keating, Dennis W. 41
Kentucky 27
King County Housing Authority (KCHA) 202
Korver-Glenn, Elizabeth 59
Krysan, Maria 57

labor market 8–9
landfills 62
Land Ordinance of 1785 44
land ownership 44
land use 40, 88; reform 144–147; and residential segregation 56, 58, 62; *see also* locally unwanted land uses (LULUs); zoning
Latino households 54
Lehman Brothers 66
LIHTC *see* Low-Income Housing Tax Credit (LIHTC) program
limited equity cooperatives 50
linkage fees 107–108, 150
loans 67, 84, 87, 129, 131; chattel 137; deferred-payment 129; forgivable 129; low-interest 129; predatory 60; *see also* mortgages
local control 43, 104; *see also* local governments
local governments: ability to meet housing needs 41; historical perspective 43–45, 47–50; policies and programs 22–23, 50–51, 71, 83–84, 87–89, 104–108, 166–167, 174–176, 186–189; and poverty 44; provision of public housing 47; regulatory strategies 144–154; role of 83–84, 87–89
Local Initiatives Support Corporation 50, 90
locally unwanted land uses (LULUs) 56
location 8–9
Los Angeles 61, 73–75, 120, 149, 196
lowest-income households 4, 9, 38–39, 48, 50, 97, 102, 113, 126; *see also* low-income households
low-income households 9; cash assistance 119–122; children from 62; dispersing 39; historical perspective 38–42, 48–51; housing cost burden 7; and housing shortage 21–23; quality 19; and wealth inequality 34; *see also* extremely low-income households; lowest-income households
Low-Income Housing Tax Credit (LIHTC) program 9–10, 21, 23, 50–51, 84, 87, 92, 102–104, 107, 151–152, 165, 197; critique of 103; income thresholds for 9; opposition to developments 61, 103
low-income neighborhoods 17, 39, 61–62
low-income renters 4, 21, 70, 109
LULUs *see* locally unwanted land uses

Mandatory Affordable Housing (MHA) program 187–188
manufactured homes 15, 136–137
Marx, Karl 37–38
matched savings programs 130

McMillan, Jimmy 4
mediation 138
mental health 5, 61, 62, 68, 71, 76, 99
Mercy Housing 91
metropolitan statistical areas (MSAs) 27, **28**, 29, 145
Miami 29
micro houses 133
Microsoft 189, 196–197
microunits 92, 133
middle-income households 9; GI Bill 60; and mortgage interest deduction 126–127; public policies 38, 39, 46, 51; and wealth inequality 34
military veterans 60, 86–87, 131
Millennials 20
Miller, Greg 48
Milwaukee 70
Minneapolis 146, 196
Minnesota 107
missing middle housing 133
Mississippi 60
mixed-income housing 41, 50, 98, 99
moderate-income households 23, 62, 86, 97, 104, 128–129, 131, 177
Moderately Priced Dwelling Unit program 136
Modern Housing proposal 41
Montgomery County 136, 203
mortality 5; and homelessness 76; infant 58
mortgages 6, 66–67; adjustable-rate (ARM) 67; and COVID-19 pandemic 69; during Great Depression 46; financing 130–131; and housing construction 21; and housing cost calculation 29; interest payments on 6–7, 67; interest tax deduction 39, 40, 42, 58, 126–127, 132, 200; loan modification 138; and people of color 17, 54, 56, 59–60; principal payments on 6; refinancing 138; shared appreciation 131; shared equity 131; subprime 67–69; *see also* foreclosure
mortgage tax credit certificate programs 132
Moving to Opportunity (MTO) 116, 160
MSAs *see* metropolitan statistical areas
multi-family housing 15, 17, 18, 41, 58, 101, 145–147, 151
Multifamily Tax Exemption (MFTE) program 187
mutual housing associations 50

National Alliance to End Homelessness 90
National Apartment Association 153
National Association of Housing and Redevelopment Officials 90
National Association of Real Estate Boards 45
National Association of Realtors 44
National Council of State Housing Agencies 90
National Fair Housing Alliance 90

National Housing Conference 90
National Housing Trust Fund 102, 196, 199
national identity 46
National Low Income Housing Coalition (NLIHC) 21, 22, 83, 90, 117, 152
Native Americans 85, 105
Native Hawaiian people 55
naturally-occurring affordable housing (NOAH) 23, 84
neighborhood-based organizations 91
neighborhoods: amenities 61, 62, 99; Black 59–61; effects 61–62; gentrifying 137; low-income 17, 39, 61–62; White 47, 56, 59; *see also* neigborhood-based organizations
NeighborWorks America 50, 90
neoliberalism 49
Netherlands 37, 38, 201
Nevada 67
New Deal 44, 48, 85, 96
New Deal Ruins 49
New Jersey 71, 153
New York 27, 45, 49, 59, 73–75, 97, 105, 133
New York City 4, 44, 61, 89, 136, 154
New York City Housing Authority (NYCHA) 89
NIMBY *see* Not In My Backyard
Nixon, Richard 49, 97
NLIHC *see* National Low Income Housing Coalition
NOAH *see* naturally-occurring affordable housing
non-profit organizations 42, 90–91
non-profit providers 49, 50, 84
Not In My Backyard (NIMBY) 58, 61, 109, 126, 148

Obama, Barack 142
Ohio 27, 29
older adults: in gentrifying neighborhoods 137; and HCV program 113, 116; and homelessness 75; and public housing 96, 101; and Supplemental Security Income (SSI) 119
online search platforms 61
Oregon 88, 147, 152
outdoor relief 44
owner occupancy 15
Own Your Home program 45

Pacific Islander people 55
PBRA *see* project-based rental assistance
PBVs *see* project-based vouchers
Pennsylvania 27
people of color 16–17; access to mortgage financing 17, 54, 56, 59–60; and evictions 54, 55; and homelessness 54, 55, 73, 75; and homeownership 17, 40, 54–56, 125–126; and housing cost burden 54; and land ownership 44
permitting process 149–150

Personal Responsibility and Work Opportunity Reconciliation Act of 1996 118, 130
PHAs *see* public housing authorities
Philadelphia 61, 117
philanthropy 91, 188–189, 196
Phillips, Shane 198, 200
physical health 61, 62, 71
place stratification 56
plumbing 19, 45
pollutants 62
poor, the, 44; *see also* poverty
poorhouses 44
population growth 35
Portland 73
poverty: cycle of 39, 99, 116, 141; deconcentrating 39, 116–117; Guideline 9; and homelessness 73–74; and local governments 44; shelter 7, 26
predatory lending 60
preference theory 55
Preserving Community and Neighborhood Choice (PCNC) 143
preterm birth 58
Private Activity Bonds 106–107
private market 15, 21, 37–39, 42, 51; housing outside of 50; and housing vouchers 61; and neoliberalism 49; *see also* private sector
private mortgage insurance (PMI) 130
private sector 84, 91–92; *see also* private market
Progressive Era 44–47
project-based rental assistance (PBRA) 100–101
project-based vouchers (PBVs) 100
Project Rental Assistance Contracts (PRAC) 100, 101
property ownership 16; *see also* homeownership
property taxes 6–7, 10, 88, 107, 108, 132, 137–138, 167, 187
property tax relief programs 137–138
property values 68, 125, 126; *see also* home values
Pruitt Igoe project 95, 97, 98, 112
public assistance 44, 120–122; *see also* housing assistance
public housing 5–6, 10, 21–22, 83, 96–100, 202–203; decline of 48–50, 97–98; historical development 41, 47–50; international comparisons 37–38, 201; and people of color 17; program 38–39, 47–50, 83, 85, 89, 95–98; rent contribution by tenants 48, 97; residents 96; *see also* public housing program
Public Housing Assessment System (PHAS) 89–90
public housing authorities (PHAs) 89–90, 96–100
Public Housing Authorities Directors Association 90
Public Housing Capital Fund 97
Public Housing Operating Fund 97
public land 44

"public neighbor" 43
public offices 126
public policy 5–6, 16, 45, 49, 198; *see also* housing policy
Puritans 43, 44

Qualified Allocation Plans (QAPs) 102

race 54–63; housing stock by 16–17; and housing tenure 17; and public policies 39; and wealth-inequality 40, 54, 62; *see also* Black people; people of color; racial discrimination
racial discrimination 16–17, 39–40, 58, 127; and homelessness 73; and residential segregation 55–57
Racially Concentrated Areas of Affluence 58
Racially/Ethnically Concentrated Areas of Poverty 58
racial segregation 48, 56–59, 159, 160
racial steering 56, 58–59
racism 16–17; and homelessness 73, 74; and housing policies 39; systemic 73, 74, 75
RAD *see* Rental Assistance Demonstration (RAD) program
Reagan, Ronald 50, 85
real estate brokers 59
redlining 17, 40, 47, 56, 59–60
Reframing Housing 176
Regional Planning Association of America 47
regulatory strategies 10, 141–154; federal 141–144; local 144–151
religious organizations 42
rent 4, 5, 6, 9; affordability periods 151–152; control 10, 88, 112, 151–154; and homelessness 74; in public housing 48, 97; regulations 151–154; stabilization 151–154; and subprime mortgage crisis 69–70; *see also* rental advertisements; rental assistance
rental advertisements 60–61
rental assistance 22, 38, 83, 112, 117; cash-based 117, 120–121; demand-side 112, 117, 120–121; project-based 100–101; supply-side 95–98, 100–101, 103
Rental Assistance Demonstration (RAD) program 100, 101
Rental Assistance Payments (RAP) 100
rental housing 15–16; international comparison 18, 37; rising costs 35; stigma associated with 16; and subprime mortgage crisis 69; types of 16; *see also* rental assistance; rental market; renter households
Rental Housing Support Program (RHSP) 167
rental market 10, 22, 29, 69; exclusionary practices in 60–61; and people of color 17; vacancy rate 161–162; *see also* rental housing
rental subsidies 10, 21–22, 76, 117, 167

Rental Supplement Program (Rent Supp) 100, 101
renter households 4, 5, 7, 10, 21, 27–31, 35–36, 70–71, 117; in Chicago 163–164; household income 35; housing costs 26–31, 35; and race 54; in San Antonio 173; in Seattle 180–182, 184; *see also* rental housing
Renters Rising 153
Rent is Too Damn High Party 4
residential density 144–148; *see also* housing density; zoning
residential mobility 56
residential segregation 55–63, 103, 142–143
residential stratification 57; *see also* residential segregation
resident-owned communities (ROCs) 136
residual approach 7
Revenue Act of 1913 126
revitalization grants 98
RHS *see* Rural Housing Services
Riis, Jacob 44
Romney, George 49
Roosevelt, Franklin D. 46, 47
Rural Housing Services (RHS) 87
Rust Belt 27, 163

St. Louis 77, 95, 98
San Antonio: background 169–170; demographic overview 170–171; federal housing supports 173–174; housing stock 171–173; proactive approach 176–177; state and local housing supports 174–176
San Antonio Housing Trust 175
San Francisco 8, 23, 26, 73, 74, 148
schools 55, 61, 62
school teachers 5
Schuetz, Jenny 194, 198, 199
Schwartz, Alex 51
Seattle 71, 73, 74, 196, 202–203; federal housing supports 186; housing stock and market 181–185; overview 179–181
Seattle Housing Authority (SHA) 186
Seattle Housing Levy 88–89, 187
Section Eight Management Assessment Program (SEMAP) 89–90
Section 8 New Construction and Substantial Rehabilitation Program (S8 NC/SR) 100
security interest 66–67
segregation 4, 39; measuring 57–58; racial 48, 56–59, 159, 160; residential 55–63, 103, 142–143; voluntary 55; *see also* desegregation
self-reliance 40
self-sufficiency 99
shared equity cooperatives 136
shared equity homeownership 135
shelter poverty 7, 26
Sherraden, Michael 130

single-family homes: average size 19–20; as the dominant form of housing 15–16, 18, 19, 145; as the focus of American housing policy 45–46; and race 17
Single-Family Housing Guaranteed Loan Program 131
single-family zoning 43, 58, 88, 144–147
single-person households 18
slums 47, 48
social housing 21, 37, 201; *see also* public housing
social isolation 61, 72
social mobility 44, 55
social networks 57, 73
social processes 57
social safety nets 38, 199–201
Social Security Disability Insurance (SSDI) 119, 121
socioeconomic status 55
Soviet Union 18
spatial assimilation 56
spatial autocorrelation 58
spatial polarization 58
SSDI *see* Social Security Disability Insurance
SSI *see* Supplemental Security Income
starter homes 20
State & City Funded Rental Housing Programs database 22
state governments 22, 49–51, 83, 105–108, 166–167, 174–177, 186–189; role of 87–89
state housing finance agencies (HFAs) 49, 51, 87
steering 56, 58–59
stewardship 135–136
Stewards of Affordable Housing for the Future 91
stigma 16, 112, 115
Stockton 120
stress 61, 68, 71
subprime mortgage crisis 67–69
subsidy programs 49
substance use 73
Supplemental Nutrition Assistance Program (SNAP) 200
Supplemental Security Income (SSI) 119, 121
supportive housing 76, 84, 86, 101
Sweden 38
Switzerland 18

TANF *see* Temporary Assistance for Needy Families
Tars, Eric 193
tax benefits 7, 29, 107, 127
tax credits 9–10, 38, 50–51, 84, 102–104, 107, 131–132; *see also* Earned Income Tax Credit (EITC); Low-Income Housing Tax Credit (LIHTC) program
Tax Cuts and Jobs Act of 2017 126

tax-exempt housing bonds 96, 106–107
tax incentive programs 107
tax increment financing (TIF) 108
Tax Reform Act of 1986 50
Temporary Assistance for Needy Families (TANF) 118, 120–122
tenant associations 91
tenant protections 10
Tenement House Act of 1901 45
Tenement House Law of 1867 44
tenements 44, 45
Terner Center for Housing Innovation 90
Texas 88, 153
third sector 135
TIF *see* tax increment financing
tiny homes 133
townhomes 15, 133
transitional housing 76
Trump, Donald 142–143

unemployment 46, 47, 69
United Kingdom 38, 201
United Nations (EU) 37
United States Housing Authority 47
upper-income households 38, 39, 200
urban ghettos 17
Urban Institute 90, 91, 199, 200
urban renewal projects 48, 49
USDA *see* U.S. Department of Agriculture
U.S. Department of Agriculture (USDA) 85, 87, 129, 131
U.S. Department of Health and Human Services 86, 130
U.S. Department of Housing and Urban Development (HUD) 6, 48, 49, 50, 58, 72, 84–85, 89, 97–101, 113–117, 142, 143, 160, 165, 174
U.S. Department of Labor 45

U.S. Department of Veterans Affairs 86–87, 131
U.S. Interagency Council on Homelessness 85, 86

vacancy: decontrol 153; rates 4, 68–69, 74, 115
vacant properties 68
Vale, Larry 37, 43
very low-income households 9, 102–103, 167, 173
Veterans Affairs Supportive Housing (VASH) 86
voluntary segregation 55

wages 5, 7, 8, 34
Wagner-Steagall Housing Act of 1937 47
Washington 88
Washington DC 71, 73, 107, 186–187, 202
wealth generation 17–18, 31, 40
wealth inequality 17, 34, 40; race-based 40, 54, 62, 127
welfare reform 118
welfare systems: limited 38; parochial approach 43–44; typology of 38
West coast cities 73–75
White households 17, 40, 47, 58, 125, 127
White neighborhoods 47, 56, 59
White people 55
women 44, 58
workhouses 44
working poor 38
World War I 45
World War II 47–48
Worst Case Housing Needs 8

Yes In My Backyard (YIMBY) program 147

Zillow (online platform) 61
zoning: exclusionary practices 133, 145–146; inclusionary 10, 88, 147–148; reform 144–147; regulations 46, 88; and residential segregation 56; single-family 43, 58, 88, 144–147

9781032407265